Water Conservation and Utilisation in Agriculture

Water Conservation and Utilisation in Agriculture

Harkanwar Singh

RANDOM PUBLICATIONS
NEW DELHI (INDIA)

Water Conservation and Utilisation in Agriculture

ISBN 978-93-5111-478-9

Published in 2015 in India by

RANDOM PUBLICATIONS

Reprint 2019

4376-A/4B, Gali Murari Lal, Ansari Road
New Delhi-110 002
Phone : +9111-43580356, 011-23289044, 011-43142548
e-mail: sales@randompublications.com,
info@randompublications.com, randomexports@gmail.com

Type Setting by : Friends Media, Delhi-110089
Printed at : Replika Press Pvt. Ltd.

Preface

As population continues to increase around the world, there is a growing demand for safe, reliable sources of water to meet the needs of the expanding population. Farmers, ranchers, and rural communities are particularly susceptible to the mounting pressures to provide more water to urban and urbanizing areas at the expense of water supplies in rural and agricultural communities. The term "agricultural water security" describes the need to maintain adequate water supplies to meet the food and fiber needs of the expanding population—maximizing the efficiency of water use by farmers, ranchers, and rural communities.

Conservation of water in the agricultural sector is essential since water is necessary for the growth of plants and crops. A depleting water table and a rise in salinity due to overuse of chemical fertilizers and pesticides has made matters serious. Various methods of water harvesting and recharging have been and are being applied all over the world to tackle the problem. In areas where rainfall is low and water is scarce, the local people have used simple techniques that are suited to their region and reduce the demand for water.

Water Conservation and Utilization in Agriculture covers various techniques of water productivity enhancement across different situations. It is a blend of applied part of agriculture technologies with sound scientific basis. It will serve a textbook for agronomy, soil science and engineering students, a reference for research scientists and teachers in the areas of crop production, integrated farming systems, dry land agriculture, cropping systems, production technology management under different situations, soil fertility management. It will also serve as a guide to extension officials of the department of agriculture and soil and water conservation engineering.

Author

Contents

1

Use of Water in Agriculture

Our Earth seems to be unique among the other known celestial bodies. It has water, which covers three-fourths of its surface and constitutes 60-70 wt % of the living world. Water regenerates and is redistributed through evaporation, making it seem endlessly renewable.

Actually, only 1% of the world's water is usable to us. About 97% is salty sea water, and 2% is frozen in glaciers and polar ice caps. Thus that 1% of the world's water supply is a precious commodity necessary for our survival. Dehydration (lack of water) will kill us faster than starvation (lack of food). Since the plants and animals we eat also depend on water, lack of it could cause both dehydration and starvation. The scenario gets worse. Water that looks drinkable can contain harmful elements, which could cause illness and death if ingested.

A water molecule contains one oxygen and two hydrogen atoms connected by covalent bonds. Thus its' chemical formula is H_2O.. Water is a liquid at temperatures above 0 °C at sea level, but it often co-exists on Earth with its solid state, ice, and gaseous state (water vapor or steam). Water also exists in a liquid crystal state near hydrophilic surfaces.

Water on Earth moves continually through the hydrological cycle of evaporation and transpiration (evapotranspiration), condensation, precipitation, and runoff, usually reaching the sea. Evaporation and transpiration contribute to the precipitation over land.Safe drinking water is essential to humans and other lifeforms. Access to safe drinking water has improved over the last decades in almost every part of the world, but approximately one billion people still lack access to safe water and over 2.5 billion lack access to adequate sanitation. There is a clear correlation between access to safe water and GDP per capita. However, some observers have estimated that by 2025 more than half of the world population will be facing water-based vulnerability. A recent report suggests that by 2030, in some developing regions of the world, water demand will exceed supply by 50%. Water plays an important role in the world economy, as it functions as a solvent for a wide variety of chemical substances and facilitates industrial cooling and transportation. Approximately 70% of the fresh water used by humans goes to agriculture.

Water Cycle

The water cycle (known scientifically as the hydrologic cycle) refers to the continuous exchange of water within the hydrosphere, between the atmosphere, soil water, surface water, groundwater, and plants.

Water moves perpetually through each of these regions in the water cycle consisting of following transfer processes:

— evaporation from oceans and other water bodies into the air and transpiration from land plants and animals into air.

— precipitation, from water vapor condensing from the air and falling to earth or ocean.

— runoff from the land usually reaching the sea.

Most water vapor over the oceans returns to the oceans, but winds carry water vapor over land at the same rate as runoff into the sea, about 47 Tt per year. Over land, evaporation and transpiration contribute another 72 Tt per year. Precipitation, at a rate of 119 Tt per year over land, has several forms: most commonly rain, snow, and hail, with some contribution from fog and dew. Condensed water in the air may also refract sunlight to produce rainbows.

Water runoff often collects over watersheds flowing into rivers. A mathematical model used to simulate river or stream flow and calculate water quality parameters is hydrological transport model. Some of water is diverted to irrigation for agriculture. Rivers and seas offer opportunity for travel and commerce. Through erosion, runoff shapes the environment creating river valleys and deltas which provide rich soil and level ground for the establishment of population centers. A flood occurs when an area of land, usually low-lying, is covered with water. It is when a river overflows its banks or flood from the sea. A drought is an extended period of months or years when a region notes a deficiency in its water supply. This occurs when a region receives consistently below average precipitation.

Water Resources

Water resources are sources of water that are useful or potentially useful. Uses of water include agricultural, industrial, household, recreational and environmental activities. Virtually all of these human uses require fresh water. 97% of the water on the Earth is salt water. However, only three percent is fresh water; slightly over two thirds of this is frozen in glaciers and polar ice caps. The remaining unfrozen freshwater is found mainly as groundwater, with only a small fraction present above ground or in the air.

Fresh water is a renewable resource, yet the world's supply of clean, fresh water is steadily decreasing. Water demand already exceeds supply in many parts of the world and as the world population continues to rise, so too does the water demand. Awareness of the global importance of preserving water for ecosystem services has only recently emerged as, during the 20th century, more than half the world's wetlands have been lost along with their valuable environmental services for Water Education. The framework for allocating water resources to water users (where such a framework exists) is known as water rights.

Sources of Fresh Water

Surface water

Surface water is water in a river, lake or fresh water wetland. Surface water is naturally replenished by precipitation and naturally lost through discharge to the oceans, evaporation, evapotranspiration and sub-surface seepage.

Although the only natural input to any surface water system is precipitation within its watershed, the total quantity of water in that system at any given time is also dependent on many other factors. These factors include storage capacity in lakes, wetlands and artificial reservoirs, the permeability of the soil beneath these storage bodies, the runoff characteristics of the land in the watershed, the timing of the precipitation and local evaporation rates. All of these factors also affect the proportions of water loss.

Human activities can have a large and sometimes devastating impact on these factors. Humans often increase storage capacity by constructing reservoirs and decrease it by draining wetlands. Humans often increase runoff quantities and velocities by paving areas and channelizing stream flow.

The total quantity of water available at any given time is an important consideration. Some human water users have an intermittent need for water. For example, many farms require large quantities of water in the spring, and no water at all in the winter. To supply such a farm with water, a surface water system may require a large storage capacity to collect water throughout the year and release it in a short period of time. Other users have a continuous need for water, such as a power plant that requires water for cooling. To supply such a power plant with water, a surface water system only needs enough storage capacity to fill in when average stream flow is below the power plant's need. Nevertheless, over the long term the average rate of precipitation within a watershed is the upper bound for average consumption of natural surface water from that watershed.

Natural surface water can be augmented by importing surface water from another watershed through a canal or pipeline. It can also be artificially augmented from any of the other sources listed here, however in practice the quantities are negligible. Humans can also cause surface water to be "lost" through pollution. Brazil is the country estimated to have the largest supply of fresh water in the world, followed by Russia and Canada.

Under River Flow

Throughout the course of a river, the total volume of water transported downstream will often be a combination of the visible free water flow together with a substantial contribution flowing through sub-surface rocks and gravels that underlie the river and its floodplain called the hyporheic zone. For many rivers in large valleys, this unseen component of flow may greatly exceed the visible flow.

The hyporheic zone often forms a dynamic interface between surface water and true ground-water receiving water from the ground water when aquifers are fully charged and contributing water to ground-water when ground waters are depleted. This is especially significant in karst areas where pot-holes and underground rivers are common.

Ground water

Sub-surface water, or groundwater, is fresh water located in the pore space of soil and rocks. It is also water that is flowing within aquifers below the water table. Sometimes it is useful to make a distinction between sub-surface water that is closely associated with surface water and deep sub-surface water in an aquifer (sometimes called "fossil water").

Sub-surface water can be thought of in the same terms as surface water: inputs, outputs and storage. The critical difference is that due to its slow rate of turnover, sub-surface water storage is generally much larger compared to inputs than it is for surface water. This difference makes it easy for humans to use sub-surface water unsustainably for a long time without severe consequences. Nevertheless, over the long term the average rate of seepage above a sub-surface water source is the upper bound for average consumption of water from that source.

The natural input to sub-surface water is seepage from surface water. The natural outputs from sub-surface water are springs and seepage to the oceans.If the surface water source is also subject to substantial evaporation, a sub-surface water source may become saline. This situation can occur naturally under endorheic bodies of water, or artificially under irrigated farmland. In coastal areas, human use of a sub-surface water source may cause the direction of seepage to ocean to reverse which can also cause soil salinization. Humans can also cause sub-surface water to be "lost" (i.e. become unusable) through pollution. Humans can increase the input to a sub-surface water source by building reservoirs or detention ponds.

Desalination

Desalination is an artificial process by which saline water (generally sea water) is converted to fresh water. The most common desalination processes are distillation and reverse osmosis. Desalination is currently expensive compared to most alternative sources of water, and only a very small fraction of total human use is satisfied by desalination. It is only economically practical for high-valued uses (such as household and industrial uses) in arid areas. The most extensive use is in the Persian Gulf.

Frozen water

Several schemes have been proposed to make use of icebergs as a water source, however to date this has only been done for novelty purposes. Glacier runoff is considered to be surface water.The Himalayas, which are often called "The Roof of the World", contain some of the most extensive and rough high altitude areas on Earth as well as the greatest area of glaciers and permafrost outside of the poles. Ten of Asia's largest rivers flow from there, and more than a billion people's livelihoods depend on them. To complicate matters, temperatures are rising more rapidly here than the global average. In Nepal the temperature has risen with 0.6 degree over the last decade, whereas the global warming has been around 0.7 over the last hundred years.

Agricultural Use of Water

The use of water for agriculture has changed the production of crops dramatically in the 20th century. Agricultural use of water accounts for nearly 70% of the water used throughout the world, and the majority of this water is used for irrigation. During the 1970s, the construction of irrigation systems dramatically increased. Its rate of growth began to decrease in both developed and developing countries in the 1980s. An increase in irrigation development guarantees an increase in crop production in many countries. Irrigation allows the land that does not recieve enough precipitation annually to become land that can be used for productive agriculture.

On the negative side, irrigation of land causes salinization of the land that is being irrigated, mostly in arid and semi-arid regions. Irrigation of cropland can increase the possibility fertilizers and pesticides will infiltrate into the groundwater or runoff into nearby streams. Along with the irrigation of crops, the farmers that have livestock must provide clean water for the livestock to drink. With a growing world population, expected to increase by 2 billion people by the year 2030, agriculture needs to find a way to use less water or to use the water more efficiently.

There are several different systems that are used for irrigation purposes, including ditch irrigation, terracing, overhead irrigation, center pivot irrigation, lateral move irrigation, and drip or trickle irrigation. Irrigation of cropland has greatly increased production of food, but has also had some drawbacks due to the amount of water that is being drawn from aquifers. Some of the problems with irrigation are competition for surface water rights, depletion of underground aquifers, ground subsidence, and buildup of toxic salts on soil surfaces in regions of high evaporation rates, called salinization. These problems can be increased or be more detrimental during periods of drought. Irrigation has been increasing between 1960 and 1995, as the graph below depicts.

Almost all of the water of the planet occurs as saltwater in the oceans. Of the 3 percent of the global resource that is freshwater, two-thirds comes as snow and ice in polar and mountainous regions. Hence, liquid freshwater constitutes about 1 percent of the global water resource. At any one time, almost all of this occurs as groundwater, while less than 2 percent of it is to be found in rivers and lakes. In the temperate, humid climates, about 40 percent of the precipitation ends up in the groundwater, and for Mediterranean-type climates the figure is 10-20 percent. For the truly dry climates, the value can be virtually zero.

Not all the water in rivers and lakes and in the groundwater is accessible for use because part of the water flows in remote rivers and during seasonal floods that cannot be captured before the water reaches the ocean. An estimated 9 000—14 000 km^3 of water is economically available each year for human use. At most, this represents 0.001 percent of the estimated global water. At present, annual withdrawals of water for human use are about 3 600 km^3. This may give the impression that there is plenty of water available that could be withdrawn for human use.

However, part of the available surface water must remain in the rivers and streams to ensure effluent dilution and safeguard the integrity of the aquatic ecosystem. How large this part should be is little understood. It varies with the time of year, and each river basin has its own specific ecological limit below which the system can be expected to degrade.

A global estimate for this demand is 2 350 km^3/year. Adding this to the amount withdrawn each year for human use results in nearly 6 000 km^3 of economically accessible water that is already committed. This indicates that globally the margin is fairly small. Because water and population are distributed unevenly throughout the world, the water situation is already critical in various countries and regions and likely to become so in several more.

Agriculture is the principal user of all water resources taken together, i.e. rainfall (so-called green water) and water in rivers, lakes and aquifers (so-called blue water). It accounts for about 70 percent of all withdrawals worldwide, with domestic use amounting to about 10 percent and industry using some 21 percent.

There is an important distinction between water withdrawn for use and water actually consumed. In irrigated agriculture, about half of the water withdrawn is consumed in evaporation and transpiration from plants and moist soil surfaces. Some of the plants that contribute to this evapotranspiration process are unproductive weeds and plants on wasteland. Water that is abstracted but not consumed infiltrates the soil and is stored as groundwater or flows back through drains into rivers. This drainage water is generally of a lower quality than the water originally withdrawn owing to contamination by agrochemicals and salts leached from the soil profile.

Compared with a return flow of 50 percent of the water withdrawn for agriculture, 90 percent of the water for domestic use is returned to rivers and aquifers as wastewater, while industry typically returns up to 95 percent. Poor-quality return flows from urban and industrial areas are sometimes treated before being returned to watercourses, but the non-point character of agricultural pollution makes treatment difficult. In this sense, agricultural water pollution may be better handled by controlling quantitative use and outflow from agricultural land.

Rainfed Crop Production

Rain is the source of water for crop production in the more humid regions of the world where some 60 percent of the world's food crops are grown. Rainfed agriculture takes place on some 80 percent of the arable land and irrigated agriculture produces 40 percent of the world's food crops on the remaining 20 percent. In order to meet future food demands, it is expected that relatively more crops will have to be grown on irrigated than on rainfed land, such that about equal amounts will come from both types of areas.

Given the importance of rainfed cereal production, insufficient attention has been paid to potential production growth in rainfed areas. Most attention usually focuses on the possible expansion of irrigated areas. However, increasing cereal yields in rainfed temperate countries, better plant protection and manure techniques, and the use of supplemental irrigation in more arid countries indicate the significant potential for improving rainfed agriculture.

In arid regions, water scarcity is the result of insufficient rainfall. Semi-arid regions may receive enough annual rainfall to support crops but the rainfall is distributed so unevenly in space and time that rainfed agriculture is barely possible. Rainfall variability generally increases with a decrease in annual amounts and it is particularly high in the Sahelian countries. These regions are known for their periodic droughts that may last several years. Rain in semi-arid regions also tends to fall in a few hard showers. Such rainfall is difficult to capture for agricultural use. This leads to large amounts of runoff going into drains and eventually seeping down to the groundwater or to rivers.

Where river discharge is large and difficult to manage, one way to capture the flow is through spate irrigation in which part or all of the river flow is diverted into fields surrounded by high bunds. In this way, one irrigation of up to 50 cm can provide enough soil moisture for a wheat crop even in the skeletal soils found in Yemen. Floodwater harvesting takes place within a streambed and entails blocking the water flow. This causes water to concentrate in the streambed. After the flood season is over, the streambed area where the water collects is cultivated. A terraced wadi (ephemeral stream) system is one type of floodwater harvesting. Here, a series of low check dams are constructed across a wadi and the wadi area is cultivated. Too much flow will breach the check dams or the diversion structures in spate irrigation. The suitability of these methods also depends on the soil conditions and depth of the wadi bed. Water harvesting, which is the collection and storage of surface runoff, has also proved useful in semi-arid regions with infrequent rains.

Although there is a great variety of such rainwater technologies, it is not clear whether their widespread use is always feasible, especially for poor farmers. The costs involved in the construction and maintenance of the water harvesting system play a major role in farmers' decisions on whether to adopt the technique or not. In the past, water harvesting systems were often installed with financial support from outside agencies, such as NGOs and international funding agencies. Many of those systems failed because of lack of involvement of the beneficiaries and their inability to organise and pay for maintenance.

Rosegrant et al. report construction costs for water harvesting systems in Turkana, Kenya, of US$625-1,015/ha. Labour and construction constitute the bulk of the water harvesting costs as the opportunity cost for using the land is essentially zero. The initial high labour costs of building the water harvesting system often provide a disincentive for adoption of the technique. Moreover, many farmers in arid or semi-arid areas do not have the human resources available to move the large amounts of earth necessary in the larger systems. Therefore, small-scale water and soil conservation techniques that are applicable at field level are often adopted more easily. Investments on a larger scale require the existence or creation of community organisations both to pay for the necessary investment and maintenance, and to manage the benefits of the water harvesting infrastructure.

Maintenance of the system is sometimes required in the rainy season when labour is relatively scarce and therefore expensive because of competition with conventional agriculture. Notwithstanding these reservations about the widespread applicability of extensive water harvesting systems, model studies indicate that there is significant scope for increasing rainfed production provided that appropriate investments and policy changes are made. Crop breeding specifically for rainfed environments is crucial to future cereal growth.

Groundwater Use for Irrigation

Groundwater use for irrigation presents a paradox: regions where the groundwater resource has been overdeveloped coexist with regions with considerable potential for development of groundwater for use in irrigated agriculture. A corollary is the so-called fallacy of aggregation: in aggregate terms, at the global or even national level, groundwater availability appears far in excess of present use. The annual groundwater use for the world as a whole has been placed at 750—800 km^3. This figure may appear modest compared to the overall groundwater reserves, but only a fraction of the world's groundwater reserves are economically available for agriculture. It is estimated that about 30 percent of the world's irrigation supply is made up by groundwater but this input accounts for some of the highest yields and highest value crops.

The number of tubewells providing water to irrigated land in India, China, Pakistan, Mexico and many other countries has grown rapidly in the past 40 years. For example, some 60 percent of irrigated cereal production in India depends on irrigation from groundwater wells. This has led to widespread and uncontrolled overabstraction of the resource and the creation of a 'bubble' groundwater economy. In Yemen, abstraction is estimated to exceed recharge by 400 percent.

Groundwater abstraction and recharge have rarely been quantified accurately. This should be a first step in assessing the potential for further development of the resource and designing management approaches. Where irrigated agriculture depends in part on pumped groundwater, many of the command areas present a mosaic of irrigation methods. These range from totally irrigated by canal water to entirely fed by pumped groundwater, with most of the fields having some combination of the two. Hence, irrigation is by definition conjunctive, but there are few examples of conjunctive management.

In China, 52 percent of the irrigated lands are (at least in part) served by tubewells. As a result of overabstraction of groundwater, water tables have fallen by up to 50 m over the last 30 years. For example, in the Fuyung Basin in north China, surface water has been curtailed drastically in order to meet industrial demand, and farmers have responded by resorting to groundwater irrigation. The root of the Asian groundwater crisis alluded to by Shah et al. that threatens millions of poor rural communities lies in the open access nature of the resource. Paradoxically, it is precisely this feature of groundwater in shallow aquifers that has made it a powerful tool in the fight against poverty, i.e. that everyone who can afford to install a pump has free access to water.

Irrigation with groundwater is generally more productive than canal irrigation because groundwater is produced close to where it is used with hardly any losses during transport. In addition, farmers are in control of the timing and amount of the water extracted. Evidence in India suggests that crop yield per cubic metre of water on groundwater-irrigated farms tends to be 1—3 times higher than on farms irrigated with surface water.

Throughout the world, most groundwater development has proceeded primarily on the basis of individual initiatives. Unlike surface irrigation or drinking-water supply projects where government agencies are generally involved in many aspects of design, financing and implementation, most groundwater development is driven by the decision of individual farmers to drill wells and buy pumps. While governments often facilitate this process through subsidies and rural electrification, large implementation departments are rare.

In consequence, there are very few government agencies that have frequent and direct contact with groundwater users. Furthermore, surface water development generally involves the diversion of flows or the construction of storage on a clearly defined stream or water body. The impact of such actions on downstream users is generally clear, at least in a conceptual sense. As a result, large bodies of customary and formal law along with the resource monitoring and enforcement systems required to implement them have developed over the long history of surface water development.

This is not generally the case with large-scale groundwater development. It is a recent phenomenon and diversions have a far less directly observable impact on other users. As a result, groundwater extraction remains highly 'individualistic' and tends to occur outside the framework of established institutions for allocating, monitoring or managing the resource base. In locations such as India, tens of millions of individuals own and operate wells. Most of these wells are on private lands. The location, use and even existence of such wells is often unknown to any individual aside from the owners and their immediate community. As a result, no established institutional basis for management exists.

Role of Surface Water

Global indicators of water scarcity tend to ignore variations in the importance of irrigated agriculture for food security among countries. They also fail to account for seasonal differences in supply. For example, more than 70 percent of the total supply in India occurs in the three monsoon months of June, July and August, when most of it floods out to the sea. Moreover, countrywide data ignore regional differences in water supply and withdrawal within the country, an example of the aggregation fallacy. Regardless of these caveats, most observers conclude that many countries do not have a surplus of water available for irrigation.

In fact, many countries do not have sufficient annual water withdrawal to irrigate their potential gross irrigated area even at high basin-irrigation efficiencies. Basin-irrigation efficiency includes all reuse of drainage water and is considerably higher than system-irrigation efficiency where drainage flow from one system is used for irrigation again downstream in another system. Most analyses indicate that, whatever the water scarcity indicator used, more than half of the world's population lives in countries with varying degrees of water scarcity. This scarcity can be physical (there is no more water), economic (the country cannot afford to develop additional water resources) or caused by a lack of social adaptive capacity. Examples of adaptive capacity are the ability to produce more value per unit water consumed, and imports 'virtual water', which is the water used to produce the crops obtained on the world market.

There is concern that more people may be affected by food insecurity as a result of water scarcity. Competition for the same resources combined with the increasing trend of water pollution exacerbates this problem. Moreover, the largely unknown impact of climate change may make water scarcity in some countries more severe. Several studies suggest that rice yields are likely to increase in the higher latitudes and decrease in the lower latitudes under future climates. It is likely that the poorest countries (and the poorest people within them) will suffer disproportionately as they are less able to adapt to the changing conditions. Future projections by the International Food Policy Research Institute (IFPRI) indicate that water withdrawals will rise by some 22 percent between 1995 and 2025. Projected withdrawals in developing countries will increase by 27 percent in the 30-year period, compared with 11 percent in developed countries. Only a very small increase in irrigated area is expected, which will be more than offset by increases in river basin efficiency.

Investments in Irrigation Infrastructure

An accurate global view of irrigation investment trends does not exist, but certain proxies can be used to indicate such trends. For example, there has been a sharp decline in World Bank lending for new irrigation schemes. Funding for new irrigation construction has largely stopped and the emphasis is on the sustainability and efficiency of existing systems. Irrigation and drainage is still one of the core investment activities of the World Bank's rural portfolio, but now mainly in support of rehabilitation and the devolution of responsibilities to water user associations (WUAs). However, the number of irrigation and drainage projects is expected to decrease to well below what it was in the 1980s.

Investments in irrigation systems are perceived to have failed to address the changing needs of irrigation services because rehabilitation of existing systems was mostly carried out to restore original project objectives. This type of rehabilitation is often inappropriate as it tends to ignore desirable changes in cropping patterns and irrigation techniques and thus allows low water-productivity practices to continue. Cost and time overruns in irrigation projects as well as public opposition to large dam building projects have further eroded the confidence of funding agents in irrigation investments.

Considering the negative aspects of irrigated agriculture (e.g. salinity, waterlogging, health hazards, and groundwater exploitation), it can no longer be taken for granted that irrigation is a protected, preferred practice and that its negative externalities will be accepted unconditionally. Nevertheless, irrigation development and dam building must continue, even if only to update existing facilities and to replace dams and reservoirs that have lost most of their storage capacity as a result of sedimentation. The loss of effective capacity of Mediterranean dams is currently between 0.5 and 1 percent/year with some as high as 3 percent. In Morocco, the reduction in regulating capacity attributable to reservoir silt-up is equivalent to a loss of irrigation potential of 6 000-8 000 ha/year. Improved erosion control in catchment areas may eventually prolong the life span of reservoirs and dams.

A reduced level of investment in irrigation may not be all bad. In the past, the construction of many irrigation systems was supply driven as part of internationally funded development aid without significant input from the future users of the scheme and sometimes against their expressed wishes. Irrigation potential was and still is seen as an important indicator for evaluating future irrigation development. This parameter expresses how much a country's irrigated area could be expanded according to land use and water availability criteria. Hence, its value changes over time depending on the country's economy and competition for water.

However, this notion of irrigation potential has often been used as the sole criterion in setting a country's agricultural and water resource policies, without a parallel analysis of economic, social, institutional and environmental constraints and without a thorough market analysis. The failure of some irrigation schemes can be attributed to a narrow focus on irrigation infrastructure and water distribution combined with an insufficient focus on the productivity of the agricultural systems and their responsiveness to agricultural markets.

The public policy of supply-driven irrigation development, adhered to by many governments and donor agencies, may be justified because of the observed importance of the role of irrigated agriculture in food security. However, the role of the private sector in irrigation development is often underestimated or ignored. The many investment decisions of smallholders and commercial farmers could exceed public investment, e.g. in Zambia and India. In particular, where there is a comparative advantage in irrigated production to service local and international markets, which may be for vegetables and cut flowers rather than traditional food crops, significant private irrigation investments appear to follow.

Irrigation and Agricultural Development

Since 1960, growth in average cereal yields has largely kept pace with the increase in world population. It is widely assumed that it will continue to do so until the population begins to stabilise. Most of the increase in grain production has been the result of yield increases rather than expansion in cropped area. Projections by FAO, the IFPRI and the World Bank assume that the further increases in cereal production will come from continuous increases in yield. However, trends in yield data collected by FAO indicate that the average world cereal yield would have to reach at least 4 tonnes/ha for a world population of 8 000 million people from its present level of about 3 tonnes/ha. At present, all the developed countries taken together have not achieved an average cereal yield of 4 tonnes/ha. This is the extent of the challenge.

The contribution from irrigated agriculture to achieving this goal will be critical as irrigation provides a powerful management tool against the vagaries of rainfall. Irrigation also makes it economically attractive to grow high-yielding crops and to apply the adequate plant nutrition and pest control required in order to obtain the full potential of these modern varieties. According to the IFPRI, while food production will increase much faster in developing countries than in developed countries, it will not keep pace with demand, and food imports will need to increase.

In 1999/2000, developing countries produced 1 030 million tonnes of grain, i.e. 55 percent of world production, and accounted for 61 percent of world grain consumption. To bridge the gap between demand and production, developing countries imported 231 million tonnes of grain, equivalent to 72 percent of worldwide imports. These statistics illustrate that developing countries play a major role in the international agricultural trade and that they are highly susceptible to changes in the world agricultural market in terms of food security. For the poorest countries, an increase in domestic agricultural production is key to improving food security. This explains why expectations about the food security role of irrigated agriculture remain high.

Agricultural development based on water conservation and irrigation is often considered a promising avenue for poverty alleviation in rural areas. For example, the availability of water for a small domestic garden plot, usually managed by women, can make a significant difference to household nutrition and thus contribute to improved livelihoods. Water harvesting may make this possible.

However, this effect is small scale and irrigated agriculture with its higher crop yields is expected to have greater impact on the incidence of poverty and malnutrition. This effect is expected regardless of whether the irrigation project is small or large scale. However, recent studies have shown that poverty alleviation as a result of irrigation development requires that the project be geared towards the needs of the poor. This includes access to training in the technical aspects of irrigation but also in community organisation and marketing.

One of the recurrent problems is the lack of access to credit, capital or land. Even microcredits have no grace period; repayments typically have to start after a few weeks. This makes them of little use for the purchase of cheap technology, such as treadle pumps and microdrip systems. It has been argued that these technologies are profitable within a short period and do not require a subsidised price for poor people or specific poverty alleviation measures. The problem of credit is not specific to irrigation development and needs to be addressed in a more general sense for successful rural development of poor regions.

Expanding irrigated areas, increasing the control of water and applying high-yield technology in irrigated agriculture have given rise to large increases in farm income, especially in Asia. However, this increase has been disproportionately in the hands of the larger peasant farmers. They are not the poorest of the poor, but their increased expenditure pattern has driven increased employment of those who are the poorest of the poor. The latter have little or no land and they benefit little even from agricultural production programmes directed most closely to them. However, they benefit from lower food prices, increased wages and growth in demand for rural non-farm goods and services. By contrast, the capital- and import-intensive consumption patterns of large-scale farmers, and especially absentee farmers, contribute much less to poverty reduction. This is more typical for some Latin American countries than for Asia and Africa.

Cost recovery from poor farmers for the operation and maintenance (O&M) of irrigation systems is controversial. Subsidising these services and providing irrigation water far below cost is financially unsustainable. Stepped tariffs in which the basic need is provided free to poor people may work in the case of drinking-water but is difficult to implement for irrigation water.

Monitoring the efficiency of water use in agriculture for many small farmers each using a small amount of water is expensive, but providing irrigation water below cost contributes to wasting of water.

In developing countries, agriculture generally produces many non-tradable goods, such as food crops of lower quality and goods with unusually high transaction costs. This aspect gives agriculture a prominent role in poverty reduction. It also buffers the national economy from shocks to international markets in agricultural commodities. For the rural poor in low-income countries, increased employment opportunities allow them to escape from poverty and hunger. Because they generally have few skills, the poor are more likely to find employment in the production of goods and services that cannot be marketed on the international market.

Examples of this type of employment include maintenance of irrigation and drainage structures, watershed management, and afforestation, and where there is a sizeable storage reservoir, employment could be found in fisheries, ecotourism and navigation. Thus, increased employment and, hence, poverty reduction depend on increased domestic demand for these non-tradable, non-farm goods and services. Agriculture is the principle source of such demand and so it is only with rising farm incomes that poverty can be reduced and food security increased.

Hence, investments in irrigation development may achieve additional goals such as enhancing economic growth and poverty alleviation in rural areas. Nonetheless, the question may be asked whether investments in other parts of the infrastructure are not more likely to achieve these goals. For example, the steady decline in poverty in India from the mid-1960s to the early 1980s was strongly associated with agricultural growth, particularly the green revolution, which coincided with massive investments in agriculture and rural infrastructure.

According to IFPRI studies in India, the impact of additional irrigation investments on poverty reduction ranked third after rural roads, and agricultural research and extension. Additional government spending on irrigation had a significant impact on productivity growth, but no discernible impact on poverty reduction. While spending on irrigation and power has been essential in the past for sustaining agricultural growth, the levels of irrigation may now be such that it may be more important to maintain rather than increase the systems. The IFPRI studies have also indicated that the marginal returns to several infrastructural investments in India are now higher in many rainfed areas. They also have a potentially greater impact on reducing rural poverty.

A global analysis of the link between farming systems and poverty indicates that the prospects for reducing agricultural poverty in the Near East and North Africa are good. However, for the region as a whole, exit from agriculture is the best household strategy for poverty reduction, followed by increased off-farm income. The study indicates that the priority roles of the State are to support the development of vital infrastructure, such as roads, water supplies, services and power supply, and to regulate resource use and pricing for water and power. By comparison, in South Asia, measures that support households in small-farm diversification and also for growth in employment opportunities in the off-farm economy were found to be the most likely to contribute to poverty reduction.

When weighing the pros and cons of new irrigation investments against the benefits of other investments, all additional potential benefits of irrigation, such as health benefits resulting from better nutrition (i.e. more calories and a more balanced diet) and greater rural employment, should be taken into account. Many of the irrigation benefits are site-specific and generalisations cannot be made. Moreover, without proper techniques for monitoring the physical performance of irrigation systems, it is impossible to assess the potential benefits that may accrue from further investments to improve them.

Notwithstanding these caveats, the fundamental question concerning the economic utility of further investments in irrigation development as a means towards rural development and poverty alleviation is an important one. At least two conclusions can be drawn from the discussion of the role of water in sustainable food production, poverty alleviation and rural development. The first is that donor agencies and governments have a difficult choice to make when investing for poverty reduction and rural development. The choice is not automatically for agriculture or water. The second conclusion is that the right set of government policies can make a large difference in food production, poverty reduction and rural development.

Agricultural Water Productivity

Productivity is a ratio between a unit of output and a unit of input. Here, the term water productivity is used exclusively to denote the amount or value of product over volume or value of water depleted or diverted. The value of the product might be expressed in different terms (biomass, grain, money). For example, the so-called 'crop per drop' approach focuses on the amount of product per unit of water. Another approach considers differences in the nutritional values of different crops, or that the same quantity of one crop feeds more people than the same quantity of another crop. When speaking of food security, it is important to account for such criteria.

Another concern is how to express the social benefit of agricultural water productivity. All the options that have been suggested can be summarised by the phrases 'nutrient per drop', 'capita per drop', 'jobs per drop', and 'sustainable livelihoods per drop'. There is no unique definition of productivity and the value considered for the numerator might depend on the focus as well as the availability of data. However, water productivity defined as kilogram per drop is a useful concept when comparing the productivity of water in different parts of the same system or river basin and also when comparing the productivity of water in agriculture with other possible uses of water.

Crop water production is governed only by transpiration. As it is difficult to separate transpiration from evaporation from the soil surface between the plants (which does not contribute directly to crop production), defining crop water productivity using evapo-transpiration rather than transpiration makes practical sense at field and system level. In irrigated agriculture in saline areas, the leaching requirement, i.e. the amount of water that needs to percolate to maintain rootzone salinity at a satisfactory level, should also be included together with evapotranspiration in the amount of water that is necessarily depleted during plant growth. Other non-productive but beneficial uses could be included. Examples are evapotranspiration by windbreaks, cover crops, and the water used in wetting seedbeds to enhance germination.

The question of considering water losses from seepage and field percolation as consumption does not receive a unique response. If this water is of no use downstream or if it generates further pollution such as that resulting from geological salt leaching, then it must be accounted for as consumption. Solutions to minimise these losses, such as canal lining or water improvement application, then have a positive effect on productivity. However, from a broader environmental point of view, it can be important to consider the impact of the outflow of an irrigation system on the overall productivity of an ecosystem.

As with the numerator, the choice of the denominator (which drops to be included) should depend on the scale, the point of view and the focus. At basin level, the choice might be between water diverted from the source and the same minus water restored, whereas at field level one might consider useful rain, irrigation water and supplemental irrigation.

Spatial Variability

Reported data on water productivity with respect to evapotranspiration (WPET) show considerable variation, e.g. wheat 0.6-1.9 kg/m^3, maize 1.2-2.3 kg/m^3, rice 0.5-1.1 kg/m^3, forage sorghum 7-8 kg/m^3 and potato tubers 6.2-11.6 kg/m^3, with incidental outliers obtained under experimental conditions. Data on field-level water productivity per unit of water applied (WPirrig), as reported in the literature, are lower than WPET and vary over an even wider range. For example, grain WPirrig for rice varied from 0.05 to 0.6 kg/m^3, for sorghum from 0.05 to 0.3 kg/m^3 and for maize from 0.2 to 0.8 kg/m^3.

The variability occurs because data were collected in different environments and under different crop management conditions. These affected the yield and the amount of water supplied. Furthermore, it is often difficult to determine the real crop yield over a large area, e.g. the size of a large irrigation system. When asked for yield figures, individual farmers are likely to give a figure that depends on the situation. For a loan application, they may overstate the yield, whereas for payment of a debt or a tariff, they will probably understate the yield obtained. Vegetable yields of vegetables may change every day, and unless good records are kept, no one will know exactly how much was harvested during the total harvest period. Yields expressed in monetary terms are more doubtful as prices on the local market may fluctuate considerably over time.

Nevertheless, water productivity data across scales are useful in assessing whether water drained from upstream is reused effectively downstream. However, there are few reliable data on water productivity at different scale levels within the same system. A study using remote sensing and GIS technologies assessed crop WPET at various irrigation system scales in the Indus Basin in Pakistan.

Crop water productivity was found to vary significantly at the scale of small canal command areas. When water productivity was aggregated for canal command areas, the highest water productivity values decreased gradually. Their variability also decreased until at a scale of about 6 million ha water productivity tended to a low value of about 0.6 kg/m^3. This arose because at the larger scale, canal commands with less fertile or saline soils and with less canal water and poorer quality groundwater were included in the average.

Increase of Water Productivity

Despite concerns about the technical inefficiency of water use in agriculture, water productivity increased by at least 100 percent between 1961 and 2001. The major factor behind this growth has been yield increase. For many crops, the yield increase has occurred without increased water consumption, and sometimes with even less water given the increase in the harvesting index. Example of crops for which water consumption experienced little if any variation during these years are rice (mostly irrigated) and wheat (mostly rainfed), for which the recorded increases worldwide amount to 100 and 160 percent respectively.

At the global level, the increase in water consumption for agriculture in the past 40 years has been 800 km^3 while world population has doubled to 6 000 million. Considering that the arable rainfed area has not increased, one can conclude that with an additional 800 km^3 of water the world has been able to feed an additional 3 000 million people. This gives a rough estimate of 0.720 m^3/d/capita. This figure is low compared to the estimated global average for 2000 of 2.4 m^3/d/capita, which includes water for food at field level not including water losses. This is a good indicator of the significant productivity gain recorded in agriculture; a gain that has enabled the world to accommodate the doubling of the population and also increase intake. As a whole, one can estimate that the water needs for food per capita halved between 1961 and 2001 from about 6 m^3/d to less than 3 m^3/d.

The importance of water needs for food makes any small relative gain in this sector equivalent to a significant gain for other uses. For example, given the water needs for capita in 2000, a 1-percent increase in water productivity in food production generates a potential of water use of 24 litres/d/capita. In order to produce the equivalent of the domestic water supply, a gain of 10 percent in agricultural water productivity would be required, which is a matter of years. Therefore, it can be argued that investing in agriculture and in agricultural water is the best avenue for freeing water for other purposes.

However, future agricultural gains will need to be split into several components:

(i) compensation for the reduction of agricultural production areas as a result of urban encroachment, soil degradation, and the depletion of water resource availability or access (groundwater);

(ii) increased water access for the rural poor and vulnerable groups;

(iii) generation of wealthier farming systems; and

(iv) freezing water for other uses including the environment.

Principles for Improving Water Productivity

The key principles for improving water productivity at field, farm and basin level, which apply regardless of whether the crop is grown under rainfed or irrigated conditions, are:

(i) increase the marketable yield of the crop for each unit of water transpired by it;

(ii) reduce all outflows (e.g. drainage, seepage and percolation), including evaporative outflows other than the crop stomatal transpiration; and

(iii) increase the effective use of rainfall, stored water, and water of marginal quality.

The first principle relates to the need to increase crop yields or values. The second one aims to decrease all 'losses' except crop transpiration. Its phrasing does not imply that it will be impossible to increase water productivity by reducing stomatal transpiration. It is conceivable that plant breeding may find ways to overcome this constraint. The third principle aims at making use of alternative water resources. The second and third principles should be considered parts of basinwide integrated water resource management (IWRM) for water productivity improvement. IWRM recognises the essential role of institutions and policies in ensuring that upstream interventions are not made at the expense of downstream water users.

These three principles apply at all scales, from plant to field and agro-ecological levels. However, options and practices associated with these principles require different approaches and technologies at different spatial scales.

Enhancing Water Productivity at Plant Level

Plant-level options rely mainly on germplasm improvements, e.g. improving seedling vigour, increasing rooting depth, increasing the harvest index (the marketable part of the plant as part of its total biomass), and enhancing photosynthetic efficiency. The most significant improvements in yield stability have usually resulted from breeding programmes to develop an appropriate growing cycle such that the duration of the vegetative and reproductive periods are well matched with the expected water supply or with the absence of crop hazards.

Planting, flowering and maturation dates are important in matching the period of maximum crop growth with the time when the saturation vapour pressure deficit is low. The periods of maximum crop growth may be optimised by means of breeding technology. Improved varieties with a deeper rooting system contribute to drought avoidance and the effective use of water stored in the soil profile. Drought escape and increasing drought tolerance are also important strategies for increasing water productivity.

Daylength-insensitive varieties of short to medium duration enabled crops, such as wheat, rice and maize varieties developed as part of the green revolution, to increase water productivity by escaping late-season drought that adversely affects flowering and grain development. The modern rice varieties have about a threefold increase in water productivity compared with traditional varieties. Progress in extending these achievements to other crops has been considerable and will probably accelerate following the recent identification of the underlying genes. Genetic engineering, if properly integrated in breeding programmes and applied in a safe manner, can further contribute to the development of drought tolerant varieties and to increasing the water use efficiency.

Raising Water Productivity

Improved practices at field level relate to changes in crop, soil and water management. They include: selecting appropriate crops and cultivars; planting methods (e.g. on raised beds); minimum tillage; timely irrigation to synchronise water application with the most sensitive growing periods; nutrient management; drip irrigation; and improved drainage for water table control.

Water depletion occurs when water evaporates from moist soil, from puddles between rows and before crop establishment. All cultural and agronomic practices that reduce these losses, such as different row spacings and the application of mulches, improve water productivity. The irrigation method also affects these evaporative losses.

Drip irrigation causes much less soil wetting than sprinkler irrigation. The significance of soil improvement in enhancing water productivity is often ignored. However, integrated crop and resource management practices, such as improved nutrient management, can increase water productivity by raising the yield proportionally more than it increases evapotranspiration. This principle applies to both irrigated and rainfed agriculture. Integrated weed and integrated pest management have also contributed effectively to yield increases.

One of the field-level methods for increasing water productivity is deficit irrigation, where deliberately less water is applied than that required to meet the full crop water demand. The prescribed water deficit should result in a small yield reduction that is less than the concomitant reduction in transpiration. Therefore, it causes a gain in water productivity per unit of water transpired. In addition, it could lower production costs if one or more irrigations could be eliminated.

For deficit irrigation to be successful, farmers need to know the deficit that can be allowed at each of the growth stages and the level of water stress that already exists in the rootzone. Most importantly, they need to have control over the timing and amount of irrigations. Deficit irrigation carries considerable risk for the farmers where water supplies are uncertain, as is the case with rainfall or unreliable irrigation supplies.

Where water availability falls below a certain level, the value of the crop can fall to zero, either because the crop dies or because the product is of such low quality as to be unmarketable. When water is scarce, farmers could reduce the irrigation as appropriate to maximise returns to water if they have control over the timing and amount of irrigations. This degree of flexibility is usually the case with sprinkler and drip irrigation, and also with pumped groundwater if the farmer owns the pump. A totally flexible delivery system for surface irrigation in large irrigation systems is expensive because of the required overcapacity in the conveyance system.

The trade-off between reduced yield and higher water productivity needs to be quantified in economic terms before recommending deficit irrigation (and other water-saving irrigations in rice production). The often cited low water productivity per unit of water supply in rice cultivation derives from considering as losses the percolation resulting from the standing water layer on the field surface. However, this water is often recycled, and rice water productivity generally compares well with that of a dry cereal. Nevertheless, water-saving irrigation techniques such as saturated soil culture and alternate wetting and drying can reduce the unproductive water outflows drastically and increase water productivity. These techniques generally lead to some yield decline in the current lowland rice high-yielding varieties.

However, some experiments are reporting substantial yield increases for local varieties using a technique called system rice intensification (SRI), a technique which originated in Madagascar. Here again there is no unique response; the fit with local resources and capacity is the most important feature to account for. Without anticipating results of current investigations in many countries, it seems that the potential of the SRI technique for the poor to increase the productivity of scarce land and water is significant provided that enough family labour is available.

Other approaches are being researched as part of efforts to increase water productivity without sacrificing yield. One of these is to develop so-called aerobic rice systems that allow rice cultivation in non-flooded conditions. The development of these new rice varieties is essential if rice is to be grown like other irrigated upland crops and the deep percolation associated with paddy rice is to be avoided.

Water-related problems in rainfed agriculture are often related to large spatial and temporal rainfall variability rather than low cumulative volumes of rainfall. The overall result of rainfall unpredictability is a high risk for meteorological droughts and intraseasonal dry spells. Bridging crop water deficits during dry spells through supplementary irrigation stabilises production and increases both production and water productivity dramatically if water is applied at the moisture-sensitive stages of plant growth.

Water harvesting for agriculture involves a storage reservoir, while in runoff farming the collected runoff is applied directly to the cultivated area. Either way, the investments in the construction of the ditches that take the runoff to the storage reservoir and of the reservoir itself are relatively small. Maintaining these structures may be more difficult if heavy rains periodically wash them away.

Many factors affect the success of rainwater harvesting. These include: the method used for runoff collection and storage; the topography; the soil characteristics (especially the infiltration rate); the choice of crop to be planted; fertiliser availability; and the effectiveness of the soil crust in the catchment area. However, probably more important than any of these physical parameters is the involvement of the beneficiaries in the design and implementation of the water harvesting structures.

Socio-economic assessments of water harvesting and supplementary irrigation are rare. It is recognised that sustainable increases in water productivity by water harvesting can only be achieved through a combination of farmer training, water conservation, supplementary irrigation, better crop selection, improved agronomic practices, and political and institutional interventions. Planning (and economic assessment) should consider explicitly the short-term effect and longer-term implications of hydrological changes brought about by water harvesting on downstream water users.

Accounting for Water Productivity

Changing the focus from the field level to system and river-basin level changes the relative importance of the various water management processes. At the larger scale, the effect of agriculture on other water users, human health and the environment becomes at least as important as production issues. Options for improving water productivity at the agro-ecological or river-basin level are found in: better land-use planning; better use of medium-term weather forecasts; improved irrigation scheduling to account for rainfall variability; and conjunctive management of various sources of water, including water of poorer quality where appropriate. Therefore, integrating germplasm improvement and resource management is crucial in the enhancement of water productivity at the field scale and above.

Gains in water productivity are possible by providing more reliable irrigation supplies, e.g. through precision technology and the introduction of on-demand delivery of irrigation supplies. However, an increase in water productivity may or may not result in greater economic or social benefits. The social benefits represent the benefits to society resulting from the water-productivity-enhancing interventions. Water in the rural areas of developing countries has many uses. Thus, water is both a public and a social good, a fact that complicates value calculations. These many uses of water include: the production of timber, firewood and fibre; and raising fish and livestock. Non-agricultural uses of water include domestic (drinking and bathing) and environmental uses.

An IWMI study of an irrigation system in Kirindi Oya in southern Sri Lanka illustrates the importance of the multiple roles of water in agriculture. The study found that at system level crops consumed only 23 percent of the total water supply, including both rainfall and external irrigation water. Of the remainder, 8 percent was used for grazing land, 6 percent evaporated from the reservoir, 16 percent was lost to the sea, 3 percent drained into lagoons, while as much as 44 percent of the water supply went to perennial vegetation that had developed since the construction of the scheme. This perennial vegetation was there because of irrigation seepage and recharge of the shallow groundwater. Tree growth is important to the people living in the area as it provides them with shade and thus improves their environment. In this project, as well as in many places in southern India, it also provides income from coconut and materials for construction (beams and ropes). Other trees are important for additional nutritional values (fruits) and some are crucial for their medicinal properties. A changeover to total control of irrigation outflow in order to increase water productivity would cause the collapse of the entire local agroforestry system.

Another example of the economic and social benefits of agroforestry is a project located along the Niger River in Mali. In this project, trees were planted on the bunds of rice fields, and also in the middle of the rice fields without affecting rice yields adversely. In this remote arid part of Mali, the value of the wooden poles of seven-year-old eucalyptus trees was so high that the farmers could pay for the O&M of the irrigation system from the sale of the trees. In another irrigation system, in southwest Burkina Faso, oil-palm and fruit trees were combined successfully with irrigated crops (mainly maize, groundnuts and industrial tomatoes). Trees were planted on ridges or on the boundaries between parcels. On the sandy, percolating soils of the irrigation system, the trees produced an important amount of complementary food and income, while the impact on the main crop was minimal.

These examples point out that not all measures to increase water productivity are appropriate in all circumstances. It is essential to consider the various uses of water in agriculture before measures are introduced that would increase water productivity at the expense of other benefits from the same source of water, especially those benefits that accrue to the local poor and landless people.

Promoting Water Productivity Gains

Using price policies to promote the economic productivity of water requires significant government intervention in order to ensure that equity of access to water and public-good issues are covered adequately. Some studies in the Indian subcontinent and elsewhere have suggested that the price for water that would be required to affect demand substantially would be about ten times the charge required to cover the O&M of the irrigation system.

A charge sufficient to cover O&M would have a minimal effect on water demand. Moreover, introducing volumetric charges for irrigation water is difficult and involves considerable expense for the installation of measuring structures and for fraud prevention. Last, in most rice-based systems in Asia, volumetric charging at individual user level or even group level is unsuitable given the permanent overflow and recycling water flows throughout the command area.

The groundwater market in India illustrates the perhaps unintended impact of government policies on the availability of water to farmers and others. Farmers in Gujarat paid about four times as much for pumped groundwater compared with farmers in Punjab and Uttar Pradesh. This difference was attributed to:

(i) differences in the way farmers were charged for the electric power to run their pumps (flat rate versus per unit consumed);

(ii) the tubewell spacing policy in Gujarat that gave each tubewell owner a monopoly over some 203 ha; and

(iii) the scarcity of public tubewells in Gujarat, which also reduced competition among groundwater suppliers.

The high prices for tubewell water in Gujarat discriminated against small and poor farmers. However, some simple changes in water policies for power pricing, tubewell spacing and public tubewells could transform groundwater markets in Gujarat into powerful instruments for small-farmer development.

Aiming for the highest economic productivity of water in agriculture may conflict with the political desire for national food security. More often than not, the economic productivity of water in growing staple crops is less than that for growing vegetables or flowers for export markets. Crop substitution involves switching high water-consuming crops for less water-consuming crops or for crops with higher economic productivity. The approach provides a strategy for increasing crop water productivity at the agro-ecological system level as well as at the global level.

Policies and incentives are important in the adoption of changes from traditional agronomic and cultural practices. However, it is necessary to identify the types of policies and incentives that will work best. Experience with conservation agriculture indicates that the short-term interests of the farmers often differ from the long-term interests of society and that the financial benefits that accrue from changes in cultural practices often take a long time to materialise. In addition, although there are large differences between individual farms, external factors also play a role, e.g. the transmission of information (via policy-related activities and social processes).

Of particular importance is the fact that the inconsistent and sometimes contradictory results from studies on the adoption of new practices suggests that the decision-making process is highly variable. This decision-making process needs to be understood more fully as it will affect the lead time from study to field practice. This lead time is often unacceptably long considering the urgent character of water-scarcity problems. Experience from participatory research and extension could help reduce this lead time.

Risk Management in Agricultural Water Use

Vulnerability to drought varies from country to country. It depends inter alia on the stage of development. Economies in the early stages of transition from subsistence farming to a more modern and productive farm economy are particularly vulnerable. This applies to much of rainfed agriculture. Rainfall patterns over Africa have not changed significantly in the past century. In particular, the Sahel, the Horn of Africa and the countries around the Kalahari Desert are characterised by high interannual and intraseasonal rainfall variability. Good and bad years do not occur at random but tend to be grouped. This has important implications for food security as food and water must be stored over a period of several poor years.

Risk is defined as the product of hazard and vulnerability. In other words, it relates to the probability of a damaging event, such as drought, and the foreseeable consequences of such an event. In terms of agriculture, the most common risk is drought. On a global scale, this risk is much greater than that of cyclones, floods and storms. However, on a regional rather than global scale, there are areas where the risk of flooding exceeds that of drought. Drought represents one of the most important natural triggers for malnutrition and famine. Drought events can be addressed at the parcel level by several management decisions, at the watershed level and at the country level. The first decisions belong to farmers or farmers collectivities whereas decisions at watershed and country level must be taken by governments or state agencies.

Risk is also defined more simply as a loss due to a damaging event. The advantage of this definition is that it can be materialised and measured easily (e.g. loss of agricultural production, loss of income). An acceptable risk is one that individuals, businesses or governments are willing to accept in return for perceived benefits. Local governments usually define the level of acceptable risk by considering information on drought hazards and combining it with economic, social and political factors specific to the area threatened.

Conflict is an ever-present risk and one of the most common causes of food insecurity. The displacement of people and the disruption of agricultural production and food distribution leaves tens of millions of people at risk of hunger and famine. Conversely, food insecurity may lead to or exacerbate conflict. Conflict in sub-Saharan Africa resulted in losses of almost US$52 000 million in agricultural output between 1970 and 1997, a sum equivalent to 75 percent of all official development assistance received by the conflict-affected countries. Conflict combined with drought has triggered six of the seven major African famines since 1980. Early warning and response can prevent famine arising from drought and other natural disasters. In war zones, lack of security and disruption of transport and social networks impede delivery of relief aid. However, several other factors contribute to food insecurity. These include: lawlessness; lack of democracy; ethnic and religious divisions; degradation or depletion of natural resources; and population pressure.

Risk Management Strategies

Stemming from the definition of risk, there are two major ways of minimising risk, either by reducing hazard or by reducing the vulnerability. Ways of minimising hazards are few and can include: rainmaking; avoidance of hail; and watershed management to content floods. Ways of minimising vulnerability can include: development of surface (including pumping water from streams) and underground irrigation facilities; integrated management of water resources; ecosystem development and diversification; education and training of farmers; early warning systems; seasonal climate forecasting; and crop insurance.

Early warning systems and seasonal weather forecasts are increasingly available to provide timely information so that governments and international aid agencies can take the necessary measures to reduce the impact of drought. However, seasonal forecasting skills are imperfect and the forecasts are not yet available to farmers. If they were, those forecasts could help farmers in choosing less water-demanding crops, e.g. sorghum rather than maize, when a drought is predicted. In the absence of reliable information about expected seasonal rainfall, some farmers will tend to accept risk in anticipation of greater profit, while others will tend to avoid risk even if there is a potential for high profit. Such risk avoidance or acceptance is a personal and a cultural characteristic.

The historical evolution of irrigated agriculture was a response to reduce the risk of crop failure in lands that were subject to periodic droughts, such as the basins of the Euphrates and Tigris rivers. Crop practices and field management provide several means for coping with soil water management. Strategies in rainfed agriculture are based on producing more food per unit of rainfall in a durable manner by collecting the maximum amount of rainfall at community, farm and parcel levels, minimising water loss at farm and parcel levels, and using water efficiently at parcel level.

The collection of maximum rainfall may involve both state and farmer organisations (water harvesting, use of recycled water from other sectors) or farmers alone (water harvesting at the farm, runoff reduction at parcel level, early planting, fallow cropping system, etc.). Minimising water loss involves farmers (evaporation reduction by mulching or rapid crop cover, windshields, minimum tillage, weeding, etc.). The efficient use of water requires farmers' involvement (use of low water consuming crop species, adapted fertilisation to available water, disease and pest control, optimal planting and seeding, selected varieties able to accomplish their cycle within the climate growing period, etc.).

Risk may be reduced substantially, while expected profit is reduced relatively less, by choosing combinations of alternatives rather than any single alternative. For example, a farmer in a rainfed area, such as Machakos in Kenya, where on average a crop of maize may yield well in one out of four years, could choose to seed one-quarter of a field with maize every year. The reality is more complicated as both the total seasonal rainfall and its distribution during the growing season have a large effect on crop yield.

The above strategies enable improved use of the available water at the parcel level. Moreover, traditional farming aims at a stable production rather than a maximum income. Farmers achieve this objective through diversification of production and low-input practices that provide that do not entail too much investment or cash. Association between farmers, for example at the village scale or within farmer groups, can further reduce the risk of low production.

Spreading Risk

Crop insurance constitutes the most explicit risk-spreading mechanism that helps distribute the cost of weather-related events through financial institutions among other economic sectors and governments. Successful examples include crop insurance for the impact of cyclones and hailstorms. The impact of drought is greater in developing countries than in developed countries, but farmers in developing countries have at best only limited access to insurance. The cost of insurance for relatively low-value staple crops is usually unaffordable.

However, spreading risk can also lead to water sharing. Water transfers within countries have occurred for some time. Some canals were constructed for navigational reasons, others to supply drinking-water to water-scarce cities, and others for agricultural purposes or various combinations of these causes. Well-known examples include the Snowy Mountain Scheme in Australia and various aqueducts in California, the United States of America. Internationally, an extensive system of link canals between the branches of the Indus River was constructed in order to ensure equity of water access between India and Pakistan following partition in 1947.

China is developing extensive water transfer schemes linking the south of the country to the water-scarce and populous north. Funding and implementing such expensive schemes in the future may help to reduce the risk of international conflict over water. Where several countries share water resources, e.g. in the river basins of the Mekong, Nile, Euphrates and Tigris, there is a perceived risk that the combination of population growth, poverty, food insecurity and water scarcity might lead to conflicts over water. Current attempts at mediation through the establishment of river basin authorities aim to reduce these risks.

Environmental Impact of Agricultural Water Use

Most production systems, agriculture included, can cause both positive and negative side-effects, or externalities that are not accounted for in markets. Agriculture's positive and negative environmental services are unintended consequences of market activities that have an impact on people other than the producer of the effect. These by-products tend not to be priced in the market and, hence, their economic values are unknown or difficult to assess.

Consideration of all the positive externalities of agriculture is not readily possible. There are cases where the same phenomenon may be positive in certain circumstances and negative in others, or it may be valued positively by some observers and negatively by others. A positive externality may reduce a negative one, and vice versa. In addition, positive and negative externalities are often linked closely, e.g. soil salinity and improved employment opportunities in irrigated agriculture.

Moreover, positive externalities are often ignored whereas negative ones tend to be reported widely. A well-known example of a negative externality is the loss of biodiversity as a result of draining wetlands for agriculture. Such losses are accelerating as human settlement continues to impinge upon wetlands and forests.

Many agricultural systems have become efficient transformers of technologies, non-renewable inputs and finance. They can produce large amounts of food, but have substantial negative impacts on capital assets. These assets comprise not only the natural resources of soil and water per se but also nutrient cycling and fixation, soil formation, biological control, carbon sequestration and pollination. The issue raises concerns about what constitutes success in agricultural production if large yield increases come at the cost of environmental and health problems. One problem is that the benefits and costs accrue to different people and are not measured in the same units.

In the 1970s and 1980s, some people considered energy to be such a common measure. Indeed, sustainable systems are much more energy efficient than modern high-input systems. Low-input rainfed rice in Bangladesh and China can produce 1.5-2.6 kg of cereal per megajoule of energy consumed. This is some 15-25 times more efficient than irrigated rice produced in Japan and the United States of America. On average, sustainable systems produce 1.4 kg of cereal per megajoule compared with 0.26 kg/MJ in conventional systems.

Modern agricultural systems depend heavily on external inputs, largely derived from fossil fuels. In most industrialised countries, energy is cheaper than labour. Hence, it seems rational to overuse natural resources and underuse labour. The result has been adverse, long-term effects on the environment. Although labour is cheaper than energy in many developing countries, agriculture often has negative effects on the environment. In relation to their policy implications, the environmental externalities of agriculture operate at different geographic scales, e.g. carbon sequestration (a positive externality) on a world scale, but salinisation of a watershed on a local scale (a negative externality).

Applying concepts such as the 'polluter pays' principle, cost recovery and cost sharing may prove unrealistic, impractical or politically disastrous to governments in countries where millions of people are poor and small-scale farmers are trying to make a living on marginal lands. A common concern in developing countries is how agricultural production in marginal areas can fulfil its primary function without depleting the natural resource base. For these reasons, developing appropriate technologies, assigning individual or common property rights, and the promotion of alternative employment outside the agricultural sector will be key strategies.

Salinity and Drainage Question

Much of the environmental impact of irrigated agriculture is linked to the management of water and salt balances of irrigated lands. This includes both minimising the amount of water required to remove salt from the root-zone, and minimising the land area required to store the salt temporarily or permanently. Good management has proved difficult. Although human-induced salinity problems can develop swiftly, solutions can be time consuming and expensive.

Various improvements in irrigation and agronomic practices can be introduced depending on the type of salinity and on the cause of the accumulation of salts to harmful levels in the rootzone. The fact that saline waters have been used successfully to grow crops shows that under some conditions, e.g. in Mediterranean climates with winter rains, saline water can be used for irrigation. Experience in other locations, where negative long-term effects from irrigating with saline or sodium-rich waters have been observed, indicates that more permanent interventions in the water and salt balance are generally required.

All arid-zone rivers have natural salt profiles, attributable to mobilisation of salts in the catchment area and saline seeps. An additional cause of river salinity is irrigation-induced transport of fossil salts owing to pumping from the groundwater into drains that discharge into the river.

Most of the drainage water from agricultural land in Punjab, Pakistan, is reused, either from surface drains or pumped up from shallow groundwater. In fact, in some systems in Punjab, one-half to two-thirds of the irrigation water is pumped from the groundwater. Therefore, the leached salts are returned to the land rather than disposed of in the river system or in evaporation ponds. The average salt influx for the Indus River water is estimated to be about twice the amount that flows out to sea. Hence, half of the annual salt influx remains in the land and the groundwater.

Most of the accumulation takes place in Punjab. A more extensive drainage system is needed in order to maintain a sustainable salt balance in the irrigated lands. Worldwide, only 22 percent of irrigated land has a drainage system (less than 1 percent of irrigated land has subsurface drainage). This makes it inevitable that more land will go out of production because of waterlogging and salinity. In general, those people who will lose their land are already very poor farmers.

The drainage situation in Pakistan is in sharp contrast with that in Egypt. In Egypt, subsurface drains that take the drainage water back to the river underlie a large portion of the irrigated land. The salts do not stay in the Nile Basin but are discharged into the Mediterranean Sea. During part of the year, the salt content in the lower Indus River is much lower than in the lower Nile River and more salt disposal into the Indus River could be accepted. However, during critically low-flow periods, such disposals would not be possible. The only option during such periods would be to store the drainage water temporarily for release during high-flood periods. Extending the Left Bank Outfall Drain, now operational in Sindh, into Punjab could provide a more permanent, but quite expensive, solution than the present inadequate number of evaporation ponds.

Wastewater Reuse

The reuse of municipal and industrial wastewater in irrigated agriculture is widespread. Some of the wastewater is treated before it is reused. However, much of it is not, and this causes significant environmental and health hazards. In addition, many of the treatment plants in developing countries operate below design capacity, which contributes to the discharge of untreated wastewater into irrigation and drainage canals. Concentrations of heavy metals in canal and drain sediments and in soil samples, as well as faecal coliform bacteria counts in canal and drainage water, often exceed WHO water-quality guidelines. For example, wastewater constitutes 75 percent of the total flow of the Bahr Bagar Drain in the Eastern Delta, Egypt, effectively turning the drain into an open sewer. Soil samples in the Eastern Delta showed cadmium levels of 5 mg/kg, more than twice the natural level.

Evidence of uptake of trace elements in crops has also been reported. For example, in the Middle Delta, Egypt, cadmium levels of 1.6 mg/kg (ppm) have been found in rice. Such levels are harmful for human health, and warrant serious attention. Thus, some of the drainage water is unfit for reuse, not because of its high salt content but because of its pollution load. In addition, safe disposal of such polluted wastewater becomes a real problem. Similar cases have been reported for other countries, e.g. Pakistan and Mexico.

The challenge of managing the conjunctive use of groundwater and canal water successfully has been alluded to before. In some areas, over-abstraction of groundwater is evidenced by the rapid dropping of water-table levels. In other areas where the groundwater is too saline for agricultural production, the water table rises as a consequence of over-irrigation and seepage from irrigation canals. Much agricultural land has gone out of production as capillary rise from shallow water tables has ruined soils and poisoned crops. Reversing this process is difficult and expensive. In India, the extent of the waterlogged areas is estimated at 6 million ha. In 12 major irrigation projects with a design command area of 11 million ha, 2 million ha are reported to be waterlogged and another 1 million ha salinised.

It is estimated that salinisation alone causes 2-3 million ha/year of potentially productive agricultural land to be taken out of production. How much of this land is reclaimed (to various degrees) and then cultivated again is unknown. Pollution of groundwater by salts and residues of agrochemicals is also a common occurrence. Where slightly saline groundwater is used for irrigation, the repeated cycles of water application to the fields, seepage of the excess water and pumping it up again from the top of the aquifer increases the salt load of the groundwater. If the vertical permeability of the aquifer is restricted, only limited mixing of seepage water takes place and the top of the aquifer from which the water is pumped becomes increasingly saline. This process has been documented for several irrigation systems in Punjab, Pakistan, where conjunctive irrigation with canal water and groundwater takes place.

The poorest farmers are those most vulnerable to environmental degradation as most of them farm under difficult growing conditions. A few farmers cultivate the best lands; the vast majority of the farmers cultivate the less fertile and marginal lands. Further degradation is likely to affect the quality of the farmers' sources of drinking-water and irrigation water, the quality of their land, possibly the quantity and quality of the fish they catch, and ultimately their health.

Lack of data on water and salt balances of irrigated land and lack of knowledge on how much water (and of what minimum quality) should be committed to downstream users frustrate attempts to allocate water more equitably to users in order to improve basin-level water productivity in agriculture. The way forward to ending unsustainable practices and reducing the concentrations of salts and agrochemicals that result directly from the degradation of the soil and water resources is a consolidated and long-term effort to improve land and water management.

Generally, agriculture and rural development have not benefited from systematic environmental analysis and management. One reason for their exclusion in the past was probably the very large number of projects (large and small) that could have been referred for an assessment, which would have overwhelmed the environmental assessment agencies. Environmental impact assessment (EIA) is usually applied to physical project planning (e.g. dams, roads, pipelines and industries), but seldom to farm practices and rural development plans. As a result, inadequate planning and inappropriate land-use practices have persisted. In many areas, soil, land and water resources are used inefficiently or are degraded, while poverty and income disparities grow.

With some 30 years of experience, EIA techniques now usually consider not only biophysical impacts but also socio-economic effects on health, human migration in and out of the project area, training of local workforce, local government capacity building, etc. Government and international policies are still needed to establish appropriate legal frameworks and an institutional base for EIA for agricultural projects. These policies should include transfer of the necessary knowledge to the rural poor, e.g. through agricultural extension services, so that they can participate in the environmental assessment of agricultural water resource management and project planning.

Modernisation of Irrigation Systems

Modernising water management in irrigation systems can be interpreted in different ways depending on the local circumstances. One type of modernisation is the introduction of modern technologies, such as water application and distribution through pipes rather than open channels, and the use of computerised soil-water sensors to trigger water applications. However, it also comprises older capital-intensive techniques, such as canal lining and land levelling. These techniques can only be introduced and used successfully where the farmers can be trained in their use or already possess the necessary skills. However, the technical side is only one aspect of modernisation. Equally important are fundamental changes in the institutional arrangements and regulations and improvements in the performance and efficacy of water users and their organisations.

Modernisation is a process of technical and managerial upgrading of irrigation schemes combined with institutional reforms, if required, with the objective to improve resource utilisation and water delivery service to farms. In this sense, modernisation offers a means of institutional reform with a purpose, not just reform for its own sake. It is systemic and practical without asking that all institutional elements change and it needs to be applied where irrigated agriculture has a clear comparative advantage.

Irrigation institutions need to adopt a service orientation and improve their performance in economic and environmental terms. This entails: adopting new technologies; modernising infrastructure; applying improved administrative principles and techniques; and promoting the participation of water users. Irrigation-sector institutions need to link their central task of providing irrigation services to agricultural production and to integrate their water demands and uses with other users at basin level. An enhanced appreciation of the water cascades and flows across landscapes and the circulation of groundwater within aquifers will lead to informed decisions on the use and reuse of agricultural water.

Because modernisation is usually perceived as an engineering project, its planning typically focuses on engineering and macroeconomic issues with only broad assumptions about how the delivery system and the on-farm irrigation systems are to be managed. Where the modernised system turns out to be incompatible with existing management practices or where unanticipated extensive changes in management practices are needed to take full advantage of the potential of the modernised system, then the modernisation project is likely to fail.In addition, failure is likely if the public irrigation organisation continues as before without the involvement and participation of the water users in the system's operation and management. Only through their involvement from the beginning of a modernisation project can farmers develop a sense of ownership and be likely to care for the system. This sense of ownership should prevent several of the problems that often arise after a short time: field channels being demolished; gates being stolen or damaged; field drainage systems becoming blocked; open drains filling with sediments and weeds; and graded land becoming spoiled by bad tillage.

An important aspect of modernisation is the effect of intended plot sizes on the feasibility of the project. For example, in Navarra, Mexico, the average plot size is about 5 000 m2 while the average area owned by one farmer is 1.3 ha. It is likely that in the near future these farms will not be economically viable for two reasons:

(i) the small plot size; and

(ii) the poor condition of the irrigation systems.

The modernisation of many irrigation systems should encompass restructuring land tenure in order to ensure plot sizes that can be farmed profitably. In this system in Mexico, such a plot size is thought to be about 5 ha. Increasing plots sizes will also allow a reduction in the investments needed to modernise the irrigation systems. In addition, farms that perform well will have the capacity to generate jobs both directly and indirectly.Nonetheless, site-specific considerations can lead to different conclusions. In Mali, the Office du Niger, which is dedicated to rice production, allocates individual plots of at least 5 ha. This allocation of large plots to full-time, maximum-profit farmers is seen to be inconsistent with the reality of people generally pursuing diversified livelihoods, especially when trying to escape from poverty. Moreover, small plots are often used more intensively. For example, in Zimbabwe in the early 1990s, the government changed its policy of giving farmers irrigated plots of 0.1 ha so they could supplement their income from rainfed agriculture, to giving each farmer 3-5 ha of irrigated land. The expectation was that the larger plot sizes would induce farmers to devote themselves full-time to irrigation. The policy also favoured giving the irrigated land to men, with the idea that they would be more likely to devote their energies to irrigated farming. However, productivity per unit of land and productivity per unit of water were later found to be higher on the smallholding system, and women farmers were significantly more likely than men to be oriented towards irrigation as their main source of food and income.

Large-scale irrigation development and modernisation projects tend to concentrate on the production of staple foods while ignoring the fisheries. A main issue with loss of fishery habitat, or specifically with the reclamation of wetlands for agriculture, is that once the wetland is converted to agricultural land, people can have title to it. Legal title to natural wetlands is not possible although traditional communal rights can be recognised. Fisheries are often taken for granted. Many people fail to see the benefits of wetlands and fisheries, whereas they realise that quantifiable benefits, such as agricultural production and hydropower, will flow from new development works.

The introduction of low-cost technologies, which could be part of the modernisation of small-scale irrigation projects, provides another example of the site-specificity of success. Inexpensive treadle pumps have been successful in some South Asian countries in extracting irrigation water from shallow aquifers. These pumps have allowed poor farmers to make good use of the available labour in their households and so increase crop production and farm income. The farmer has full control over the timing and amount of this pumped water, which given the effort involved is used sparingly.

References

Molden, D. (Ed). *Water for food, Water for life: A Comprehensive Assessment of Water Management in Agriculture.* Earthscan/IWMI, 2007.

Chartres, C. and Varma, S. *Out of water. From Abundance to Scarcity and How to Solve the World's Water Problems* FT Press (USA), 2010.

Pearce, Fred *When the Rivers Run Dry: Water—The Defining Crisis of the Twenty-First Century* Beacon Press, 2006.

The World Bank. *Sustaining Water for All in a Changing Climate*. The International Bank for Reconstruction and Development. 2010.

2

Importance of Water Conservation

With 70% of the earth's surface being water, it would seem logical to think there is plenty for all of our needs. We live in a world of water, but approximately 97% of it makes up the oceans. Ocean water is too salty to be used for drinking water, farming, or manufacturing. Only 3% of the world's water is fresh, and 2% of this supply is frozen in glaciers and ice caps. How much water does this leave us for use in everyday life? You got it, only 1%.

The demands on our supply of water increase every year. The challenge of today is to learn how to use our water wisely. This challenge is greater now than ever before as industry and population continue to grow. The United States has always had a plentiful and easily available water supply. Water has been cheap and unfortunately people have been careless and wasteful. They have dumped untreated sewage, farm chemicals, and other wastes into rivers and lakes, spoiling the water. It is necessary to to start conserving our water because the supply of cheap, easily available water is shrinking, and the development of new supplies will become more costly. If we hope to keep costs down we must all conserve.

Whether you live in Australia or China or the US, it is time to wake up and take responsiblity. It is easy to practice water conservation in the home, but there is more to be done. Our world needs help on a commercial level as well so that our waste can be controlled in such areas as agriculture and irrigation. Demands are increasing every year for water while resources are becoming more and more limited. Since many individuals are unaware (or, sadly, just don't care) that this issue needs attention, it is up to more informed and proactive individuals and companies to take up the slack.

A 40% increase is expected in water demand over the period of next two decades. The increase in water demand is a contribution of various factors including growing population, increased agricultural needs, industrial use of water and water needed for electricity production.

The problem of water waste is severe in countries where people are using the same inefficient methods for irrigation of agricultural land. Water needs are increasing every year and the proven fact is that clean water is not available to 1 out of 5 people on earth.

We use water in so many ways in our daily life. Water is a substance which covers ¾ part of the world. It means water is occupying more portion compared to land. But this water is becoming more polluted because of the environmental changes. So it is necessary to conserve the water. It is every one duty to protect the water because everyone uses this water and we the people are making water polluted. Now we may get the doubt how to conserve the water. A few simple changes in home about the water consumption make a vast contribution to water conservation.

The basic changes we have to do is stop leaking of water. We have to see whether there is a leakage in our taps. Because of the water leakage taps we can't protect the water. We can also protect the water byInstalling Low-Flow Showerheads. With these we can protect the water because it flows lower. And we won't use much water with these Low-Flow Showerheads. When we are brushing our teeth or when we are shaving we have to turn off faucets. After completing the work we have to use the water. By this change also we can preserve the water. Using sprinklers for yard is also a good remedy for water conservation. Turn off the water in the shower when shampooing the hair Water will be saved. it is better to replace the new toilets in place of old toilets. These are high efficiency and will use less water. To collect a rain water use a rain barrel at the bottom of gutter down spouts. It may use for watering plants. These are some of the things to follow to conserve the water.

We all use the water we must also see how we are using this. It is important to note that there are many reasons why we waste this natural resource; it has become necessary to waste the resources for innovating new technologies. How ever we have now proved it is not necessary to waste. It is easy now to conserve the water. By conserving water we can also save the money. There are hundred of ways to conserve the water. But may not follow all those ways. We can follow only those ways which are easier to us. The easier things we can follow are the things which are mentioned above. They are the things we follow easily in our home.

It is the responsibility of every one to educate our children. With this we are indirectly conserving the water in future also. The children who are educating to save the water will save the water in future also. Now if we cut the usage of water it definitely helps the future generations. Water is mainly related to environment. Green also helps for the environment. Growing greenery every where and encouraging forest causes rain fall. So green is also conserving the water.

Goals of Water Conservation

Water conservation can be defined as:

— Any beneficial reduction in water loss, use or waste as well as the preservation of water quality.

— A reduction in water use accomplished by implementation of water conservation or water efficiency measures; or,

— Improved water management practices that reduce or enhance the beneficial use of water.

The goals of water conservation efforts include as follows:

— *Sustainability.* To ensure availability for future generations, the withdrawal of fresh water from an ecosystem should not exceed its natural replacement rate.

— *Energy conservation.* Water pumping, delivery, and wastewater treatment facilities consume a significant amount of energy. In some regions of the world over 15% of total electricity consumption is devoted to water management.

— *Habitat conservation.* Minimizing human water use helps to preserve fresh water habitats for local wildlife and migrating waterfowl, as well as reducing the need to build new dams and other water diversion infrastructures.

Water conservation programs are typically initiated at the local level, by either municipal water utilities or regional governments. Common strategies include public outreach campaigns, tiered water rates (charging progressively higher prices as water use increases), or restrictions on outdoor water use such as lawn watering and car washing. Cities in dry climates often require or encourage the installation of xeriscaping or natural landscaping in new homes to reduce outdoor water usage.

One fundamental conservation goal is universal metering. The prevalence of residential water metering varies significantly worldwide. Recent studies have estimated that water supplies are metered in less than 30% of UK households, and about 61% of urban Canadian homes (as of 2001). Although individual water meters have often been considered impractical in homes with private wells or in multifamily buildings, the U.S. Environmental Protection Agency estimates that metering alone can reduce consumption by 20 to 40 percent. In addition to raising consumer awareness of their water use, metering is also an important way to identify and localize water leakage. Water metering would benefit society in the long run it is proven that water metering increases the efficiency of the entire water system, as well as help unnecessary expenses for individuals for years to come. One would be unable to waste water unless they are willing to pay the extra charges, this way the water department would be able to monitor water usage by public, domestic and manufacturing services.

Some researchers have suggested that water conservation efforts should be primarily directed at farmers, in light of the fact that crop irrigation accounts for 70% of the world's fresh water use. The agricultural sector of most countries is important both economically and politically, and water subsidies are common. Conservation advocates have urged removal of all subsidies to force farmers to grow more water-efficient crops and adopt less wasteful irrigation techniques.

New technology poses a few new options for consumers, features such and full flush and half flush when using a toilet are trying to make a difference in water consumption and waste. Also available in our modern world is shower heads that help reduce wasting water, old shower heads are said to use 5-10 gallons per minute. All new fixtures available are said to use 2.5 gallons per minute and offer equal water coverage.

Water Conservation Practices

Conservation is a supplemental or even an alternative technology for meeting safe drinking water needs. Conservation implemented as part of a long-term strategy for providing safe and reliable drinking water. One of chief purposes of conservation is to avoid, postpone, or reduce capital costs associated with new facilities.

Water conservation reduces demand for water through improvements in efficiency and diminishing water waste. A carefully planned and implemented long-term water conservation program can reduce water consumption 10-20 percent over 10-20 years. In addition, when droughts occur, short-term demand reduction measures – typically rationing, water use restrictions, and restrictive pricing schemes – can be implemented in addition to public information campaigns to achieve further reductions in water use. It is incumbent upon agency planners particularly within water scarce regions to have an appropriate framework for incorporating water conservation into the integrated water resources planning process.

Water users can be divided into two basic groups: system users (such as residential users, industries, and farmers) and system operators (such as municipalities, state and local governments, and privately owned suppliers). These users can choose from among many different water use efficiency practices, which fall into two categories:

1. Engineering practices: practices based on modifications in plumbing, fixtures, or water supply operating procedures
2. Behavioral practices: practices based on changing water use habits

Engineering Practices

Plumbing

An engineering practice for individual residential water users is the installation of indoor plumbing fixtures that save water or the replacement of existing plumbing equipment with equipment that uses less water. Low-flow plumbing fixtures and retrofit programmes are permanent, one-time conservation measures that can be implemented automatically with little or no additional cost over their life times. In some cases, they can even save the resident money over the long term.

Low-Flush Toilets. Residential demands account for about three-fourths of the total urban water demand. Indoor use accounts for roughly 60 percent of all residential use, and of this, toilets (at 3.5 gallons per flush) use nearly 40 percent. Toilets, showers, and faucets combined represent two-thirds of all indoor water use. More than 4.8 billion gallons of water is flushed down toilets each day in the United States. The average American uses about 9,000 gallons of water to flush 230 gallons of waste down the toilet per year. In new construction and building rehabilitation or remodeling there is a great potential to reduce water consumption by installing low-flush toilets.

Conventional toilets use 3.5 to 5 gallons or more of water per flush, but low-flush toilets use only 1.6 gallons of water or less. Since low-flush toilets use less water, they also reduce the volume of wastewater produced. Effective January 1, 1994, the Energy Policy Act of 1992 requires that all new toilets produced for home use must operate on 1.6 gallons per flush or less. Toilets that operate on 3.5 gallons per flush will continue to be manufactured, but their use will be allowed for only certain commercial applications through January l, 1997.

Even in existing residences, replacement of conventional toilets with low-flush toilets is a practical and economical alternative. The effectiveness of low-flush toilets has been demonstrated in a study in the City of San Pablo, California. In a 30-year-old apartment building, conventional toilets that used about 4.5 gallons per flush were replaced with low-flush toilets that use approximately 1.6 gallons per flush. The change resulted in a decrease in water consumption from approximately 225 gallons per day per average household of 3® persons to 148 gallons per day per household a savings of 34 percent! Although the total cost for replacement of the conventional toilets with low-flush toilets was about $250 per unit (including installation), the water conservation fixtures saved an average of $46 per year from each unit's water bill. Therefore, the cost for the replacement of the conventional toilet with a low-flush toilet could be recovered in 5.4 years.

Toilet Displacement Devices. Plastic containers (such as plastic milk jugs) can be filled with water or pebbles and placed in a toilet tank to reduce the amount of water used per flush. By placing one to three such containers in the tank (making sure that they do not interfere with the flushing mechanisms or the flow of water), more than l gallon of water can be saved per flush. A toilet dam, which holds back a reservoir of water when the toilet is flushed, can also be used instead of a plastic container to save water. Toilet dams result in a savings of 1 to 2 gallons of water per flush.

Low-Flow Showerheads. Showers account for about 20 percent of total indoor water use. By replacing standard 4.5-gallon-per-minute showerheads with 2.5-gallon-per-minute heads, which cost less than $5 each, a family of four can save approximately 20,000 gallons of water per year. Although individual preferences determine optimal shower flow rates, properly designed low-flow showerheads are available to provide the quality of service found in higher-volume models.

Faucet Aerators. Faucet aerators, which break the flowing water into fine droplets and entrain air while maintaining wetting effectiveness, are inexpensive devices that can be installed in sinks to reduce water use. Aerators can be easily installed and can reduce the water use at a faucet by as much as 60 percent while still maintaining a strong flow. More efficient kitchen and bathroom faucets that use only 2 gallons of water per minute—unlike standard faucets, which use 3 to 5 gallons per minute—are also available.

Pressure Reduction. Because flow rate is related to pressure, the maximum water flow from a fixture operating on a fixed setting can be reduced if the water pressure is reduced. For example, a reduction in pressure from 100 pounds per square inch to 50 psi at an outlet can result in a water flow reduction of about one-third. Homeowners can reduce the water pressure in a home by installing pressure-reducing valves. The use of such valves might be one way to decrease water consumption in homes that are served by municipal water systems. For homes served by wells, reducing the system pressure can save both water and energy. Many water use fixtures in a home, however, such as washing machines and toilets, operate on a controlled amount of water, so a reduction in water pressure would have little effect on water use at those locations.

A reduction in water pressure can save water in other ways: it can reduce the likelihood of leaking water pipes, leaking water heaters, and dripping faucets. It can also help reduce dishwasher and washing machine noise and breakdowns in a plumbing system. Water use in homes was compared among different water pressure zones throughout the city. Elevation of a home with respect to the elevation of a pumping station and the proximity of the home to the pumping station determine the pressure of water delivered to each home. Homes with high water pressure were compared to homes with low water pressure. An annual water savings of about 6 percent was shown for homes that received water service at lower pressures when compared to homes that received water services at higher pressures.

Gray Water Use. Domestic wastewater composed of wash water from kitchen sinks and tubs, clothes washers, and laundry tubs is called gray water.

Gray water can be used by homeowners for home gardening, lawn maintenance, landscaping, and other innovative uses. The City of St. Petersburg, Florida, has implemented an urban dual distribution system for reclaimed water for nonpotable uses. This system provides reclaimed water for more than 7,000 residential homes and businesses.

Landscaping

Lawn and landscape maintenance often requires large amounts of water, particularly in areas with low rainfall. Outdoor residential water use varies greatly depending on geographic location and season. On an annual average basis, outdoor water use in the arid West and Southwest is much greater than that in the East or Midwest. Nationally, lawn care accounts for about 32 percent of the total residential outdoor use. Other outdoor uses include washing automobiles, maintaining swimming pools, and cleaning sidewalks and driveways.

Landscape Irrigation. One method of water conservation in landscaping uses plants that need little water, thereby saving not only water but labor and fertilizer as well. A similar method is grouping plants with similar water needs. Scheduling lawn irrigation for specific early morning or evening hours can reduce water wasted due to evaporation during daylight hours. Another water use efficiency practice that can be applied to residential landscape irrigation is the use of cycle irrigation methods to improve penetration and reduce runoff. Cycle irrigation provides the right amount of water at the right time and place, for optimal growth. Other practices include the use of low-precipitation-rate sprinklers that have better distribution uniformity, bubbler/ soaker systems, or drip irrigation systems.

Xeriscape Landscapes. Careful design of landscapes could significantly reduce water usage nationwide. Xeriscape landscaping is an innovative, comprehensive approach to landscaping for water conservation and pollution prevention. Traditional landscapes might incorporate one or two principles of water conservation, but xeriscape landscaping uses all of the following: planning and design, soil analysis, selection of suitable plants, practical turf areas, efficient irrigation, use of mulches, and appropriate maintenance.

Benefits of xeriscape landscaping include reduced water use, decreased energy use (less pumping and treatment required), reduced heating and cooling costs because of carefully placed trees, decreased storm water and irrigation runoff, fewer yard wastes, increased habitat for plants and animals, and lower labor and maintenance costs.

More than 40 states have initiated xeriscape projects. Some communities use contests and demonstration gardens to promote public awareness. El Paso Water Utilities and the Council of El Paso Garden Clubs sponsor an annual "Accent Sun Country" contest. The contest spotlights homes that have water-conserving landscapes consisting of plants and grasses that require only a minimum of supplemental water and yet beautify the homes. The winning entries are publicised, and cash prizes are awarded. People are invited to tour the grounds to get ideas on how they, too, can save water, time, and money while maintaining an attractive landscape. The offices of the Southwest Florida Water Management District in Tampa and Brooksville offer free xeriscape tours every month. The tours begin with a slide show on the principles of xeriscape and continue with a walking tour of water-saving landscaping.

Behavioral Practices

Behavioral practices involve changing water use habits so that water is used more efficiently, thus reducing the overall water consumption in a home. These practices require a change in behavior, not modifications in the existing plumbing or fixtures in a home. Behavioral practices for residential water users can be applied both indoors in the kitchen, bathroom, and laundry room and outdoors. In the kitchen, for example, 10 to 20 gallons of water a day can be saved by running the dishwasher only when it is full. If dishes are washed by hand, water can be saved by filling the sink or a dishpan with water rather than running the water continuously. An open conventional faucet lets about 5 gallons of water flow every 2 minutes.

Water can be saved in the bathroom by turning off the faucet while brushing teeth or shaving. Water can be saved by taking short showers rather than long showers or baths and turning the water off while soaping. This water savings can be increased even further by installing low-flow showerheads, as discussed earlier. Toilets should be used only to carry away sanitary waste. Households with lead-based solder in pipes that flush the first several gallons of water should collect this water for alternative nonpotable uses (e.g., plant watering).

Water can be saved in the laundry room by adjusting water levels in the washing machine to match the size of the load. If the washing machine does not have a variable load control, water can be saved by running the machine only when it is full. If washing is done by hand, the water should not be left running. A laundry tub should be filled with water, and the wash and rinse water should be reused as much as possible.

Outdoor water use can be reduced by watering the lawn early in the morning or late in the evening and on cooler days, when possible, to reduce evaporation. Allowing the grass to grow slightly taller will reduce water loss by providing more ground shade for the roots and by promoting water retention in the soil. Growing plants that are suited to the area ("indigenous" plants) can save more than 50 percent of the water normally used to care for outdoor plants.

As much as 150 gallons of water can be saved when washing a car by turning the hose off between rinses. The car should be washed on the lawn if possible to reduce runoff. Additional savings of water can result from sweeping sidewalks and driveways instead of hosing them down. Washing a sidewalk or driveway with a hose uses about 50 gallons of water every 5 minutes. If a home has an outdoor pool, water can be saved by covering the pool when it is not in use.

Industrial/commercial users can apply a number of conservation and water use efficiency practices. Some of these practices can also be applied by users in the other water use categories.

Water Reuse and Recycling

Water reuse is the use of wastewater or reclaimed water from one application such as municipal wastewater treatment for another application such as landscape watering. The reused water must be used for a beneficial purpose and in accordance with applicable rules (such as local ordinances governing water reuse). Some potential applications for the reuse of wastewater or reclaimed water include other industrial uses, landscape irrigation, agricultural irrigation, aesthetic uses such as fountains, and fire protection. Factors that should be considered in an industrial water reuse programme include:

- Identification of water reuse opportunities
- Determination of the minimum water quality needed for the given use
- Identification of wastewater sources that satisfy the water quality requirements
- Determination of how the water can be transported to the new use

The reuse of wastewater or reclaimed water is beneficial because it reduces the demands on available surface and ground waters. Perhaps the greatest benefit of establishing water reuse programmes is their contribution in delaying or eliminating the need to expand potable water supply and treatment facilities. Water recycling is the reuse of water for the same application for which it was originally used. Recycled water might require treatment before it can be used again. Factors that should be considered in a water recycling programme include:

— Identification of water reuse opportunities
— Evaluation of the minimum water quality needed for a particular use
— Evaluation of water quality degradation resulting from the use
— Determination of the treatment steps, if any, that might be required to prepare the water for recycling

Use of Water for Cooling

The use of water for cooling in industrial applications represents one of the largest water uses in the United States. Water is typically used to cool heat-generating equipment or to condense gases in a thermodynamic cycle. The most water-intensive cooling method used in industrial applications is once-through cooling, in which water contacts and lowers the temperature of a heat source and then is discharged. Recycling water with a recirculating cooling system can greatly reduce water use by using the same water to perform several cooling operations. The water savings are sufficiently substantial to result in overall cost savings to the industry. Three cooling water conservation approaches that can be used to reduce water use are evaporative cooling, ozonation, and air heat exchange.

In industrial/commerical evaporative cooling systems, water loses heat when a portion of it is evaporated. Water is lost from evaporative cooling towers as the result of evaporation, drift, and blowdown. (Blowdown is a process in which some of the poor-quality recirculating water is discharged from the tower in order to reduce the total dissolved solids.) Water savings associated with the use of evaporative cooling towers can be increased by reducing blowdown or water discharges from cooling towers.

The use of ozone to treat cooling water (ozonation) can result in a five-fold reduction in blowdown when compared to traditional chemical treatments and should be considered as an option for increasing water savings in a cooling tower. Air heat exchange works on the same principle as a car's radiator. In an air heat exchanger, a fan blows air past finned tubes carrying the recirculating cooling water. Air heat exchangers involve no water loss, but they can be relatively expensive when compared with cooling towers.

The Pacific Power and Light Company's Wyodak Generating Station in Wyoming decided to use dry cooling to eliminate water losses from cooling-water blowdown, evaporation, and drift. The station was equipped with the first air-cooled condenser in the western hemisphere. Steam from the turbine is distributed through overhead pipes to finned carbon steel tubes. These are grouped in rectangular bundles and installed in A-frame modules above 69 circulating fans. The fans force some 45 million cubic feet per minute (ft3/min) of air through 8 million square feet of finned-tube surface, condensing the steam.

Rinsing

Another common use of water by industry is the application of deionized water for removing contaminants from products and equipment. Deionized water contains no ions (such as salts), which tend to corrode or deposit onto metals. Historically, industries have used deionized water excessively to provide maximum assurance against contaminated products. The use of deionized water can be reduced without affecting production quality by eliminating some plenum flushes (a rinsing procedure that discharges deionized water from the rim of a flowing bath to remove contaminants from the sides and bottom of the bath), converting from a continuous-flow to an intermittent-flow system, and improving control of the use of deionized water.

Deionized water can be recycled after its first use, but the treatment for recycling can include many of the processes required to produce deionized water from municipal water. The reuse of once-used deionized water for a different application should also be considered by industry, where applicable, because deionized water is often more pure after its initial use than municipal water.

Landscape Irrigation

Another way that industrial/commercial facilities can reduce water use is through the implementation of efficient landscape irrigation practices. There are several general ways that water can be more efficiently used for landscape irrigation, including the design of landscapes for low maintenance and low water requirements, the use of water-efficient irrigation equipment such as drip systems or deep root systems, the proper maintenance of irrigation equipment to ensure that it is working properly, the distribution of irrigation equipment to make sure that water is dispensed evenly over areas where it is needed, and the scheduling of irrigation to ensure maximum water use.

Water-saving irrigation practices fall into three categories: field practices, management strategies, and system modifications. Field practices are techniques that keep water in the field, distribute water more efficiently across the field, or encourage the retention of soil moisture. Examples of these practices include the chiseling of extremely compacted soils, furrow diking to prevent runoff, and leveling of the land to distribute water more evenly. Typically, field practices are not very costly.

Management strategies involve monitoring soil and water conditions and collecting information on water use and efficiency. The information helps in making decisions about scheduling applications or improving the efficiency of the irrigation system. The methods include measuring rainfall, determining soil moisture, checking pumping plant efficiency, and scheduling irrigation.

System modifications require making changes to an existing irrigation system or replacing an existing system with a new one. Because system modifications require the purchase of equipment, they are usually more expensive than field practices and management strategies. Typical system modifications include adding drop tubes to a center pivot system, retrofitting a well with a smaller pump, installing surge irrigation, or constructing a tailwater recovery system.

Agricultural Irrigation

Agricultural irrigation represents approximately 40 percent of the total water demand nationwide. Given that high demand, significant water conservation benefits could result from irrigating with reused or recycled water. Water reuse is the use of wastewater or reclaimed water from one application for another application. Reused water must be used for a beneficial purpose and in accordance with applicable rules. Water recycling is the reuse of water for the same application for which it was originally intended.

Factors that should be considered in an agricultural water reuse programme include:

— The identification of water reuse opportunities
— Determination of the minimum water quality needed for the given use
— Identification of wastewater sources that satisfy the water quality requirements
— Determination of how the water can be transported to the new use

Water reuse for irrigation is already in widespread use in rural areas and is also applicable in areas where agricultural sites are near urban areas and can easily be integrated with urban reuse applications.

Behavioral Practices of Agricultural Water Users

Behavioral practices involve changing water use habits to achieve more efficient use of water. Behavioral practices for agricultural water users can be applied to irrigation application rates and timing. Changes in water use behavior can be implemented without modifying existing equipment. For example, better irrigation scheduling can result in a reduction in the amount of water that is required to irrigate a crop effectively. The careful choice of irrigation application rates and timing can help farmers to maintain yields with less water. In making scheduling decisions, irrigators should consider:

— The uncertainty of rainfall and crop water demand
— The limited water storage capacity of many irrigated soils
— The limited pumping capacity of irrigation systems
— Rising pumping costs as a result of higher energy prices

Metering

Metering. The measurement of water use with a meter provides essential data for charging fees based on actual customer use. Billing customers based on their actual water use has been found to contribute directly to water conservation. Meters also aid in detecting leaks throughout a water system. In 1977, for example, Boston, Massachusetts, could not account for the use of 50 percent of the water in its municipal water system. After installing meters, the city identified leaks and undertook a vigorous leak detection programme. Unaccounted-for water dropped to 36 percent after metering and leak detection programmes were started.

Submetering. Submetering is used in units such as apartments, condominiums, and trailer homes to indicate water use by those individual units; the entire complex of units is metered by the main supplier. Submetering of water use in apartment or business complexes makes it possible to bill tenants for the water that they actually use rather than for a percentage of the total water use for the complex. Submetering makes water users more aware of how much water they use and its cost, and tenants who conserve water can benefit from lower water use costs. Submetering is reported to reduce water usage by 20 to 40 percent.

Leak Detection

One way to detect leaks is to use listening equipment to survey the distribution system, identify leak sounds, and pinpoint the exact locations of hidden underground leaks. As mentioned in the previous section, metering can also be used to help detect leaks in a system. An effective way to conserve water is to detect and repair leaks in municipal water systems. Repairing leaks controls the loss of water that water agencies have paid to obtain, treat, and pressurize. The early detection of leaks also reduces the chances that leaks will cause major property damage.

When water leaks from a system before it reaches the consumer, water agencies lose revenue and incur unnecessary costs. Such costs should provide an incentive for system operators to implement a leak detection programme. Programmes for finding and repairing leaks in water mains and laterals (conduits) might be cost-effective in spite of their high initial costs. Leak detection programmes have been especially important in cities that have large, old, deteriorating systems.

Water Main Rehabilitation

A water utility can improve the management and rehabilitation of a water distribution network by using a distribution system database. Using the database can help to lower maintenance costs and can result in more efficient use of the water resource. The database can help the utility manager to set priorities and efficiently allocate rehabilitation funds. A comprehensive database should include information on the following:

— The characteristics of the system's components, such as size, age, and material

— The condition of mains, such as corrosion

— Soil conditions or type
— Failure and leak records
— Water quality
— High/low pressure problems
— Operating records, such as pump and valve operations
— Customer complaints
— Meter data
— Operating and rehabilitation costs

Water Reuse

Another practice that should be considered by water system operators who operate publicly owned treatment works is the reuse of treated wastewater. As discussed earlier, water reuse is the use of wastewater or reclaimed water from one application for another application. Some potential applications for water reuse include landscape irrigation, agricultural irrigation, aesthetic uses such as fountains, industrial uses, and fire protection. These factors should be considered in a water reuse programme:

— The identification of water reuse opportunities
— The determination of the minimum water quality needed for the given use
— The identification of wastewater sources that satisfy the water quality requirements
— The determination of how the water can be transported to the new use.

Well Capping

Well capping is the capping of abandoned artesian wells whose rusted casings spill water in a constant flow into drainage ditches. In Seminole County, Florida, state hydrologists estimate that 1,500 abandoned artesian wells are discharging 54 Mgal/d. To put that in perspective, utilities in Seminole County pump less than 40 Mgal/d. The cost to plug such wells is about $750 (1990 dollars) per well. The state legislature has required that all such wells be capped beginning in 1993.

Pricing

Information and education promoting conservation do not appear to be effective by themselves in achieving a conservation goal without at the same time imposing significant price increases to provide a financial incentive to conserve water. Customers use less water when they have to pay more for it and use more when they know they can afford it. However, most people consider water to be a "free good" and are not willing to pay higher prices that reflect the true costs associated with the water delivered to their homes. Rate structures have the advantage of avoiding the costs of overt regulation, restrictions, and policing while retaining a greater degree of individual freedom of choice for water customers.

Overall charges for water service increased at an average compound rate of 7 percent per year during the 1980s nearly double the rate of inflation. There is concern over "price gouging" due to increased water rates. Some pricing has been objected to on the grounds that it can lead to a substantial excess of revenues over costs an excess that might be inequitable and, in some states, unconstitutional. Water utility managers must establish and design water rates that meet revenue requirements and are fair and equitable to all customer classes in the water system. This task involves the following procedures:

— Determination of the water utility's total annual revenue requirements for the period for which the rates are to be in effect
— Determination of service costs by allocation of the total annual revenue requirements to the basic water system cost components and distribution of these costs to the various customer classes in accordance with their service requirements
— Design of water rates to recover the cost of service from each class of customer

Several price rate structuring alternatives are available for water system operators. Increasing Block Rate, or Tiered, Pricing. Increasing block rate, or tiered, pricing reduces water use by increasing per-unit charges for water as the amount used increases.

Decreasing Block Rate Pricing. Decreasing block rate prices reflect per-unit costs of production and delivery that go down as customers consume more water. The monthly water use records of 101 customers were measured in a study of municipal water use in the city of Denton, Texas. Summer water use records from 1976 to 1980 during a decreasing block rate period were compared to summer use records from 1981 to 1985 during an increasing block rate period. It was found that the decreasing block rate scenario encouraged greater water use, whereas the increasing block rate scenario resulted in a reaction to the price increase and a corresponding decrease in water use.

Time-of-Day Pricing. Time-of-day pricing charges users relatively higher prices during a utility's peak use periods. Because customers are sensitive to price increases, these charges curtail demand. Time-of-day pricing can cut generating capacity and reduce reliance on expensive secondary fuel sources.

Water Surcharges. A water surcharge imposes a higher rate on excessive water use. The customer pays more money per gallon for water use that is considered higher-than-average. Surcharges include unit surcharges, winter/ summer ratios, and alternative seasonal rates. The unit surcharge method establishes a threshold level for excess consumption based on average daily per capita or per-household consumption. A surcharge is imposed for all water use above that threshold level.

For the winter/summer ratio, metered water use during the winter period is compared to consumption during the corresponding summer period, and a higher rate or surcharge is imposed for water consumption above the average winter use. Typically, an increase in usage of 14-20 percent occurs during the summer. Under an alternative seasonal rate structure, all water used during the summer or peak season is billed at a higher rate than that used during the other seasons. The increased rate is applied to all customers at all water-use levels.

Retrofit Programmes

Retrofit programmes are another tool system operators can use to promote water use efficiency practices. Retrofitting involves the replacement of existing plumbing equipment with equipment that uses less water. The most successful water-saving fixtures are those which operate in the same manner as the fixtures they are replacing--for example, toilet tank inserts, shower flow restrictors, and low-flow showerheads. A retrofit programme can involve the use of education programmes to let users know which fixtures are best, where to get them, and how to install them. System operators can also purchase water-efficient fixtures and resell them at cost to the users, but the most successful retrofit programmes have been those in which the system operator purchases, distributes, and installs the fixtures.

Retrofit programmes have been shown to be cost-effective and useful in conserving water in many cases. An apartment building in New England with 151 units was retrofitted with low-flow showerheads and faucet aerators at a cost of $1,074. As a result of the retrofit 1,725,000 gallons of water, $8,590 for energy, and $980 for water were saved in 1 year. In another retrofit programme, the Lower Colorado River Authority installed low-flow showerheads and toilet dams in an apartment complex and public housing programme in Marble Falls, Texas. Indoor per capita water use was reduced by 21 percent (from 81 to 64 gal/cap/day) in the apartment complex and was reduced 11 percent (from 102 to 91 gal/cap/day) in the public housing programme.

Current use of low-flow toilets throughout Texas could reduce the need to build new water and wastewater treatment plants by 15 percent, resulting in a savings of as much as $3.4 billion during the next 50 years. Residential water and sewer bills could also be reduced by as much as $200 million over the long term. The Texas Water Development Board estimates that the use of water-efficient plumbing fixtures should save a typical four-member household 55,800 gallons of water and $627 in reduced water and energy costs per year. The Board estimates that the use of low-flow fixtures might reduce water use statewide by 805 Mgal/d by the year 2040.

Retrofit programmes can be combined with water audit programmes to further improve potential water savings.

Residential Water Audit Programmes

Residential water audit programmes involve sending trained water auditors to participating family homes, free of charge, to encourage water conservation efforts. Auditors visit participating homes to identify water conservation opportunities, such as repairing leaks and low-flow plumbing, and to recommend changes in water use practices to reduce home water use. The audit programmes try to stretch existing water supplies by getting water users to use water more efficiently. The largest percentage of indoor use comes from bathing and toilet flushing. Therefore, the bathroom is an ideal place for water system operators to focus water conservation efforts.

Public Education Programmes

Public education programmes can be used to inform the public about the basics of water use efficiency:

— How water is delivered to them
— The costs of water service
— Why water conservation is important
— How they can participate in conservation efforts

Public education is an essential component of a successful water conservation programme. A number of tools can be used to educate the public: bill inserts, feature articles and announcements in the news media, workshops, booklets, posters and bumper stickers, and the distribution of water-saving devices. Public school education is also an important means for instilling water conservation awareness. Another way to provide public information and education, as well as to collect real-world data on water conservation and use efficiency, is through the use of demonstration projects.

A study of water demand in the United States using American Water Works Association (AWWA) data indicated that water users are more sensitive to a change in price in the South and the West than in the other regions of the country. Public education appears to have reduced water usage in the West. A heightened awareness of water's scarcity might make educational programmes more effective in the West than in the rest of the country.

Index of Water Efficiency

An index of water efficiency, or "W-Index," can be used as a device to evaluate residential water savings and as a way to motivate water users to adopt water-saving practices. A W-Index can serve as a measure of the effectiveness of water efficiency features in a home. The index provides a calculated numerical value for each dwelling unit, which is derived from the number and kind of water-saving features present, including indoor and outdoor water savers and water harvesting or recycling systems. Architects, builders, appraisers, homeowners, water suppliers, or water management agencies can use the W-Index as a basis for evaluating the water-saving capability of any particular single- or multi-family dwelling unit. Typically, an overall W-Index rating of W-50 would be considered fair, W-80 good, and W-110 excellent, based on a specific set of community water conservation goals.

Planning for Resource Protection

Monitoring and managing land use and waste disposal practices around water supply sources can potentially reduce the need for new water supply development and keep water treatment costs to a minimum. Adverse effects on a water supply source can be lessened through land use controls such as land preservation, nonregulatory and regulatory watershed programmes, environmental assessment requirements, and zoning. The protection of a water source by a utility can range from simple sanitary surveys of a watershed to the development and implementation of complex land use controls.

Water supply source protection should play an important role in the overall management of a municipal water utility. Contamination of a water source can result from point and nonpoint sources of pollution such as chemical spills, waste discharges, or the improper use and runoff of insecticides and herbicides. The contamination of a water supply source can result in the need to develop expensive treatment systems or to find new sources for water supply.

Drought Management Planning

When less rain falls than usual, there is less water to maintain normal soil moisture, stream flows, and reservoir levels and to recharge ground water. Falling levels of surface waters create unattractive areas of exposed shoreline and reduce the capacity of surface waters to dilute and carry municipal and industrial wastewater. Water quality often decreases as water quantity decreases, adversely affecting fish and wildlife habitats. In addition, dry conditions make trees more prone to insect damage and disease and increase the potential for grass and forest fires.

A drought management plan should address a range of issues, from political and technical matters to public involvement. Managing a resource essential to people's welfare during disaster and dealing with the associated emotional, economic, and physical consequences makes drought management a very challenging task.

Agricultural Applications of Water Conservation

For crop irrigation, optimal water efficiency means minimizing losses due to evaporation, runoff or subsurface drainage while maximizing production. An evaporation pan in combination with specific crop correction factors can be used to determine how much water is needed to satisfy plant requirements. Flood irrigation, the oldest and most common type, is often very uneven in distribution, as parts of a field may receive excess water in order to deliver sufficient quantities to other parts. Overhead irrigation, using center-pivot or lateral-moving sprinklers, has the potential for a much more equal and controlled distribution pattern. Drip irrigation is the most expensive and least-used type, but offers the ability to deliver water to plant roots with minimal losses. However, drip irrigation is increasingly affordable, especially for the home gardener and in light of rising water rates. There are also cheap effective methods similar to drip irrigation such as the use of soaking hoses that can even be submerged in the growing medium to eliminate evaporation.

As changing irrigation systems can be a costly undertaking, conservation efforts often concentrate on maximizing the efficiency of the existing system. This may include chiseling compacted soils, creating furrow dikes to prevent runoff, and using soil moisture and rainfall sensors to optimize irrigation schedules. Usually large gains in efficiency are possible through measurement and more effective management of the existing irrigation system.

References

EPA. "How to Conserve Water and Use It Effectively". Washington, DC. Retrieved 2010-02-03.

Geerts, S.; Raes, D. "Deficit irrigation as an on-farm strategy to maximize crop water productivity in dry areas". *Agric. Water Manage* 96 (9): 1275–1284. 2009.

Pimentel, Berger, *et al.* "Water resources: agricultural and environmental issues". *BioScience* 54 (10): 909. 2004.

U.S. Environmental Protection Agency (EPA) Cases in Water Conservation (Report). Retrieved 2010-02-02.

Vickers, Amy (2002). *Water Use and Conservation.* Amherst, MA: water plow Press. 2002.

3

Rainwater Harvesting

Water harvesting means capturing rain where it falls or capturing the run off in your own village or town. And taking measures to keep that water clean by not allowing polluting activities to take place in the catchment. Therefore, water harvesting can be undertaken through a variety of ways:

1. Capturing runoff from rooftops
2. Capturing runoff from local catchments
3. Capturing seasonal floodwaters from local streams
4. Conserving water through watershed management

These techniques can serve the following the following purposes:

1. Provide drinking water
2. Provide irrigation water
3. Increase groundwater recharge
4. Reduce stormwater discharges, urban floods and overloading of sewage treatment plants
5. Reduce seawater ingress in coastal areas.

In general, water harvesting is the activity of direct collection of rainwater. The rainwater collected can be stored for direct use or can be recharged into the groundwater. Rain is the first form of water that we know in the hydrological cycle, hence is a primary source of water for us. Rivers, lakes and groundwater are all secondary sources of water. In present times, we depend entirely on such secondary sources of water. In the process, it is forgotten that rain is the ultimate source that feeds all these secondary sources and remain ignorant of its value. Water harvesting means to understand the value of rain, and to make optimum use of the rainwater at the place where it falls.

History

Rainwater harvesting and utilisation systems have been used since ancient times and evidence of roof catchment systems date back to early Roman times. Roman villas and even whole cities were designed to take advantage of rainwater as the principal water source for drinking and domestic purposes since at least 2000 B.C. In the Negev desert in Israel, tanks for storing runoff from hillsides for both domestic and agricultural purposes have allowed habitation and cultivation in areas with as little as 100mm of rain per year. The earliest known evidence of the use of the technology in Africa comes from northern Egypt, where tanks ranging from 200-2000m^3 have been used for at least 2000 years – many are still operational today.

The technology also has a long history in Asia, where rainwater collection practices have been traced back almost 2000 years in Thailand. The small-scale collection of rainwater from the eaves of roofs or via simple gutters into traditional jars and pots has been practiced in Africa and Asia for thousands of years. In many remote rural areas, this is still the method used today. The world's largest rainwater tank is probably the Yerebatan Sarayi in Istanbul, Turkey. This was constructed during the rule of Caesar Justinian (A.D. 527-565). It measures 140m by 70m and has a capacity of 80,000 cubic metres.

Advantages

Rainwater harvesting systems can provide water at or near the point where water is needed or used. The systems can be both owner and utility operated and managed. Rainwater collected using existing structures (i.e., rooftops, parking lots, playgrounds, parks, ponds, flood plains, etc.), has few negative environmental impacts compared to other technologies for water resources development. Rainwater is relatively clean and the quality is usually acceptable for many purposes with little or even no treatment. The physical and chemical properties of rainwater are usually superior to sources of groundwater that may have been subjected to contamination. Some other advantages of rainwater harvesting include:

i) Rainwater harvesting can co-exist with and provide a good supplement to other water sources and utility systems, thus relieving pressure on other water sources.

ii) Rainwater harvesting provides a water supply buffer for use in times of emergency or breakdown of the public water supply systems, particularly during natural disasters.

iii) Rainwater harvesting can reduce storm drainage load and flooding in city streets.

iv) Users of rainwater are usually the owners who operate and manage the catchment system, hence, they are more likely to exercise water conservation because they know how much water is in storage and they will try to prevent the storage tank from drying up.

v) Rainwater harvesting technologies are flexible and can be built to meet almost any requirements. Construction, operation, and maintenance are not labour intensive.

Forms of Water Harvesting

Various forms of water harvesting (WH) have been used traditionally throughout the centuries. Some of the very earliest agriculture, in the Middle East, was based on techniques such as diversion of "wadi" flow onto agricultural fields. In the Negev Desert of Israel, WH systems dating back 4000 years or more have been discovered. These schemes involved the clearing of hillsides from vegetation to increase runoff, which was then directed to fields on the plains.

Floodwater farming has been practised in the desert areas of Arizona and northwest New Mexico for at least the last 1000 years. The Hopi Indians on the Colorado Plateau, cultivate fields situated at the mouth of ephemeral streams. Where the streams fan out, these fields are called "Akchin". Pacey and Cullis describe microcatchment techniques for tree growing, used in southern Tunisia, which were discovered in the nineteenth century by travellers.

In the "Khadin" system of India, floodwater is impounded behind earth bunds, and crops then planted into the residual moisture when the water infiltrates. The importance of traditional, small scale systems of WH in Sub-Saharan Africa is just beginning to be recognized. Simple stone lines are used, for example, in some West African countries, notably Burkina Faso, and earth bunding systems are found in Eastern Sudan and the Central Rangelands of Somalia.

Water harvesting technology is especially relevant to the semi-arid and arid areas where the problems of environmental degradation, drought and population pressures are most evident. It is an important component of the package of remedies for these problem zones, and there is no doubt that implementation of WH techniques will expand.

Traditional water management techniques most of them being simple, sure to implement and of low capital investment. The classical sources of irrigation water are often at the break of overuse and therefore untapped sources of (irrigation) water have to be sought for increasing agricultural productivity and providing sustained economic base. Water harvesting for dry-land agriculture is a traditional water management technology to ease future water scarcity in many arid and semi-arid regions of world. This old technology is gaining new popularity these days. As the appropriate choice of technique depends on the amount of rainfall and its distribution, land topography, soil type and soil depth and local socio-economic factors, these systems tend to be very site specific.

The water harvesting methods applied strongly depend on local conditions and include such widely differing practices as bunding, pitting, microcatchments water harvesting, flood water and ground water harvesting. There are three types of water harvesting techniques

Rainwater Harvesting

Rainwater harvesting is defined as a method for inducing, collecting, storing and conserving local surface runoff for agriculture in arid and semi-arid regions. Three types of water harvesting are covered by rainwater harvesting. a) Water collected from roof tops, courtyards and similar compacted or treated surfaces is used for domestic purpose or garden crops. b) Micro-catchment water harvesting is a method of collecting surface runoff from a small catchment area and storing it in the root zone of an adjacent infiltration basin. The basin is planted with a tree, a bush or with annual crops. c) Macro-catchment water harvesting, also called harvesting from external catchments, is the case where runoff from hill-slope catchments is conveyed to the cropping area located at hill foot on flat terrain.

Flood Water Harvesting

Flood water harvesting can be defined as the collection and storage of creek flow for irrigation use. Flood water harvesting, also known as 'large catchment water harvesting' or 'Spate Irrigation', may be classified into following two forms: a) In case of 'floodwater harvesting within stream bed', the water flow is dammed and as a result, inundates the valley bottom of the flood plain. The water is forced to infiltrate and the wetted area can be used for agriculture or pasture improvement. b) In case of 'floodwater diversion', the wadi water is forced to leave its natural course and conveyed to nearby cropping fields.

Groundwater Harvesting

Groundwater harvesting is a rather new term and employed to cover traditional as well as unconventional ways of ground water extraction. Qanat systems, underground dams and special types of wells are few examples of the groundwater harvesting techniques. For example, Qanats, widely used in Iran, Pakistan, North Africa and even in Spain, consists of a horizontal tunnel that taps underground water in an alluvial fan, brings it to the surface due to gravitational effect. Qanat tunnels have an inclination of 1-2% and a length of up to 30 km. Many are still maintained and deliver steadily water to fields for agriculture production and villages for drinking water supply.

Groundwater dams like 'Subsurface Dams' and 'Sand Storage Dams' are other fine examples of groundwater harvesting. They obstruct the flow of ephemeral streams in a river bed; the water is stored in the sediment below ground surface and can be used for aquifer recharge. Sand filled reservoirs have the following advantages:

i) Evaporation losses are reduced,

ii) no reduction in storage volume due to siltation,

iii) stored water is less susceptible to pollution, and

iv) health hazards due to mosquito breeding are avoided.

In most rain and floodwater harvesting schemes the water delivered by surface runoff and overland flow is stored (only) in the soil matrix. This means, that its application is limited to the rainy season. To allow cropping outside the rainy season, a number of storage media are employed, ranging from ferrocement tanks of a few m^2 content to large reservoirs, storing millions of m^3. In India and Sri Lanka, more than 500000 tanks store rain water, sometimes supplemented by water from streams or small rivers.

Tanks play several important roles e.g. as flood-control system and in preventing soil erosion and wastage of runoff during periods of heavy rainfall. Additionally, they recharge the groundwater in surrounding areas. The larger ones, 10 to 30 hectares in size, feed several thousand hectares of irrigated land. They are equipped with sluices, which deliver water to an extensive canal system. Without this tank system, paddy cultivation in large parts of the country would be impossible. These rainwater reservoirs are not only employed for irrigation in arid or semi-arid regions, but in semi-humid areas (up to 1300 mm/a rainfall), too.

As several disadvantages are connected with surface storage of water - large evaporation losses, loss of storage caused by siltation, pollution problems and loss of agricultural land, underground storage of water may be an interesting alternative. This storage can be done in near surface aquifers (e.g. in wadi beds), calling for a conjunctive management of water resources, or in cisterns. Cisterns are man-made caves or underground constructions to store water. Often the walls of these cistern are plastered; their water losses by deep percolation or by evaporation can be minimal. The construction of cisterns was already practised several thousand years ago; chalky rocks were preferred. Traditionally, in Mediterranean houses, one cellar room was specifically designed to store rainwater. Similar in-house cisterns are known from Rajasthan, NW India. In the same region, 'Kunds', covered underground tanks with a plastered catchment, are found. Nowadays cisterns are often constructed using concrete.

Identification of Rain and Flood Harvesting Areas

The most important parameters to be considered in identifying areas suitable for rain and floodwater harvesting are as follows:

a) *Rainfall*: The knowledge of rainfall characteristics (intensity and distribution) for a given area is one of the pre-requisites for designing a water harvesting system. The availability of rainfall data series in space and time and rainfall distribution are important for rainfall-runoff process and also for determination of available soil moisture. A threshold rainfall events (e.g. of 5 mm/event) is used in many rainfall runoff models as a start value for runoff to occur. The intensity of rainfall is a good indicator of which rainfall is likely to produce runoff. Useful rainfall factors for the design of a rain- or floodwater harvesting system include:

i) Number of days in which the rain exceeds the threshold rainfall of the catchment, on a weekly or monthly basis.
ii) Probability and occurrence for the mean monthly rainfall.
iii) Probability and reoccurrence for the minimum and maximum monthly rainfall.
iv) Frequency distribution of storms of different specific intensities.

b) *Land use or vegetation cover*: Vegetation is another important parameter that affects the surface runoff. In West Africa and Syria proved that an increase in the vegetation density results in a corresponding increase in interception losses, retention and infiltration rates which consequently decrease the volume of runoff. Vegetation density can be characterised by the size of the area covered under vegetation. There is a high degree of congruence between density of vegetation and suitability of the soil to be used for cropping.

c) *Topography and terrain profile*: The land form along with slope gradient and relief intensity are other parameters to determine the type of water harvesting. The terrain analysis can be used for determination of the length of slope, a parameter regarded of very high importance for the suitability of an area for macro-catchment water harvesting. With a given inclination, the runoff volume increases with the length of slope. The slope length can be used to determine the suitability for macro or micro- or mixed water harvesting systems decision making.

d) *Soil type & soil depth*: The suitability of a certain area either as catchment or as cropping area in water harvesting depend strongly on its soils characteristics viz.

i) surface structure; which influence the rainfall-runoff process,
ii) the infiltration and percolation rate; which determine water movement into the soil and within the soil matrix, and
iii) the soil depth incl. soil texture; which determines the quantity of water which can be stored in the soil.

e) *Hydrology and water resources*: The hydrological processes relevant to water harvesting practices are those involved in the production, flow and storage of runoff from rainfall within a particular project area. The rain falling on a particular catchment area can be effective (as direct runoff) or ineffective (as evaporation, deep percolation). The quantity of rainfall which produces runoff is a good indicator of the suitability of the area for water harvesting.

f) *Socio-economic and infrastructure conditions*: The socio-economic conditions of a region being considered for any water harvesting scheme are very important for planning, designing and implementation. The chances for success are much greater if resource users and community groups are involved from early planning stage onwards. The farming systems of the community, the financial capabilities of the average farmer, the cultural behaviour together with religious belief of the people, attitude of farmers towards the introduction of new farming methods, the farmers knowledge about irrigated agriculture, land tenure and property rights and the role of women and minorities in the communities are crucial issues.

g) *Environmental and ecological impacts*: Dry area ecosystems are generally fragile and have a limited capacity to adjust to change. If the use of natural resources (land and water), is suddenly changed by water harvesting, the environmental consequences are often far greater than foreseen. Consideration should be given to the possible effect on natural wetlands as on other water users, both in terms of water quality and quantity.

New water harvesting systems may intercept runoff at the upstream part of the catchment, thus depriving potential down stream users of their share of the resources. Water harvesting technology should be seen as one component of a regional water management improvement project. Components of such integrated plans should be the improvement of agronomic practices, including the use of good plant material, plant protection measures and soil fertility management.

Water Harvesting in Arid and Semi-arid Regions

Water harvesting is applied in arid and semi-arid regions where rainfall is either not sufficient to sustain a good crop and pasture growth or where, due to the erratic nature of precipitation, the risk of crop failure is very high.

Water harvesting can significantly increase plant production in drought prone areas by concentrating the rainfall/runoff in parts of the total area. The intermittent character of rainfall and runoff and the ephemerality of floodwater flow requires some kind of storage. There might be some kind of interim storage in tanks, cisterns or reservoirs or soil itself serves as a reservoir for a certain period of time.

Water harvesting is based on the utilisation of surface runoff; therefore it requires runoff producing and runoff receiving areas. In most cases, with the exception of floodwater harvesting from far away catchments, water harvesting utilises the rainfall from the same location or region. Water harvesting projects are generally local and small scale projects.

The central elements of all water harvesting techniques are: a runoff area (catchment) with a sufficiently high run-off coefficient (impermeability would be optimal), and a "run-on" area, where the accumulated water is stored and/ or utilised. In most cases the runoff is used for agricultural crops, the water then being stored in the soil profile. A high storage capacity of the soil (i.e. medium textured soils) and a sufficient soil depth (> 1 m) are prerequisites here. The water retention capacity has to be high enough to supply the crops with water until the next rainfall event.

The most important parameters to be taken into consideration in practising water harvesting are therefore: rainfall distribution, rainfall intensity, runoff characteristics of the catchment, water storage capacity of soils, cisterns or reservoirs, the agricultural crops, available technologies and socio-economic conditions.

Water harvesting played a more important role in the past for the well-being of people in dry areas than it currently does. The reasons are manyfold:

i) no pumping from groundwater or other deep water sources
ii) very few large dams
iii) no long distance conveying of water through lined canals, pipes etc.

The building of structures for water harvesting, the cleaning and smoothing of runoff surfaces, the maintenance of canals and reservoirs etc. are labour demanding: Labour was a cheap resource, or even unpaid as in the case of slaves.

Agriculture was the backbone of the society and very few other choices to generate income were given. Therefore, relatively more input was invested in agriculture including runoff agriculture. Various examples shall be given to illustrate the past role of water harvesting worldwide.

In Jordan, there is indication of early water harvesting structures believed to have been constructed over 9,000 years ago. Evidence exists that simple water harvesting structures were used in Southern Mesopotamia as early as in 4,500 BC. Internationally, the most widely known runoff-irrigation systems have been found in the semi-arid to arid Negev desert region of Israel. Runoff agriculture in this region can be traced back as far as the 10th century BC when it was introduced by the Israelites of that period.

The Negev's most productive period in history however, began with the arrival of the Nabateans late in the 3rd century B.C. Runoff farming continued throughout Roman rule and reached its peak during the Byzantine era.

In North Yemen, a system dating back to at least 1,000 B.C. diverted enough floodwater to irrigate 20,000 hectares (50,000 acres) producing agricultural products that may have fed as many as 300,000 people. Farmers in this same area are still irrigating with floodwater, making the region perhaps one of the few places on earth where runoff agriculture has been continuously used since the earliest settlement.

In the South Tihama of Saudi Arabia, flood irrigation is traditionally used for sorghum production. Today, approximately 35,000 ha land, supporting 8,500 to 10,000 farm holdings, are still being flood irrigated.

In Baluchistan two water harvesting techniques were already applied in ancient times: the "Khuskaba" system and the "Sailaba" system. The first one employs bunds being built across the slope of the land to increase infiltration. The latter one utilises floods in natural water courses which are captured by earthen bunds.

In India, the "tank" system is traditionally the backbone of agricultural production in arid and semi-arid areas. The tanks collect rainwater and are constructed either by bunding or by excavating the ground. It is estimated that 4 to 10 hectares of catchment are required to fill one hectare of tank bed. In West Rajastan, with desert-like conditions having only 167 mm annual precipitation, large bunds were constructed as early as the 15th century to accumulate runoff. These "Khadin" create a reservoir which can be emptied at the end of the monsoon season to cultivate wheat and chickpeas with the remaining moisture.

A similar system called "Ahar" developed in the state of Bihar. Ahars are often built in series. It was observed that brackish groundwater in the neighbourhood of Ahars became potable after the Ahar was built, due to increased supply of rain water.

A very old flood diversion technique called "warping" is found in China's loess areas which harvests water as well as sediment.

Since at least Roman times water harvesting techniques were applied extensively in North Africa. Archeological research by the UNESCO Libyan Valleys team revealed that the wealth of the "granary of the Roman empire" was largely based on runoff irrigation. The team excavated structures in an area several hundred kilometres from the coast in the Libyan pre-desert, where the mean annual precipitation is well below fifty millimetres. The farming system here lasted well over 400 years and it sustained a large stationary population, often wealthy, which created enough crops to generate even a surplus. It produced barley, wheat, olive oil, grapes, figs, dates, sheep, cattle and pigs.

The precipitation is variable, falling in just one or two rain storms, often separated by droughts several years long. In Algeria, the "lacs collinaires", the rainwater storage ponds are traditional means of water harvesting for agriculture. The open ponds are mainly used for watering animals. In Tunesia, the "Meskat" and the "Jessour" systems have a long tradition, but are also still practised. The "Meskat" microcatchment system consists of an impluvium called "meskat", of about 500 m^2 in size, and a "manka" or cropping area of about 250 m^2. Thus, the CCR is 2:1. Both are surrounded by a 20 cm high bund, equipped with spillways to let runoff flow into the "manka" plots.

Whereas the "Meskats" are mainly found in the Sousse region, the "Jessour" are widespread in the South. The "Jessour" system is a terraced wadi system with earth dikes ("tabia") which are often reinforced by dry stone walls ("sirra"). The sediments accumulating behind the dikes are used for cropping.

In Lybia, archeological and historical studies have revealed the development and expansion of a highly successful dry (runoff based) farming agriculture during Roman times. On the slopes of the western and eastern mountain ranges some of these techniques continue to be practised.

In Egypt, the North-West coast and the Northern Sinai areas have a long tradition in water harvesting. Remnants from Roman times are frequently found. Some wadi terracing structures have been in use for over centuries.

Traditional techniques of water harvesting have been reported from many regions of Sub Saharan Africa. The central rangelands of Somalia are home to two small scale water harvesting systems which have been important local components of the production system for generations: The Caag system is a technique used to impound runoff from small water courses, gullies or even roadside drains. Sometimes ditches are dug to direct water into the fields. Runoff is impounded by the use of earth bunds. The entire plot may be a hectare or more in size. The alignment of the bunds is achieved by eye and by experience. In this system, runoff is impounded to a maximum depth of 30 cm. If water stands for more than five days or so, the bund may be deliberately breached to prevent waterlogging

The Mossi in Burkina Faso also constructed rock bunds and stone terraces in the past. The "Zay" system in Burkina Faso is a form of pitting which consists of digging holes that have a depth of 5 - 15 cm and a diameter of 10 - 30 cm. The usual spacing is between 50 - 100 cm. This results in a CCR of about 1 - 3:1. Manure and grasses are mixed with some of the soil and put into the zay. The rest of the soil is used to form a small dike downslope of the pit. Zay are applied in combination with bunds to conserve runoff, which is slowed down by the bunds. Many other traditional water harvesting systems existed or still exist, but the basic problem is that knowledge and information in this zone is extremely limited and fragmentary.

During this century, only very few water harvesting activities in research or implementation were undertaken before 1950. Australian farmers had already started to harvest water for domestic and animal use after World War I. During World War II, some water harvesting activities were carried out on islands with high rainfall.

In Jordan, earth dams have been constructed since 1964 in order to force runoff to infiltrate for pasture improvement. At the final stage the total area flooded shall be about 2,500 ha. In 1972, a project known as "Jordan Highland Development Project" was initiated. Rock dams, contour stone bunds, trapezoidal bunds and earth contour bunds are used to increase soil moisture around the trees planted on steep lands. The total area utilised since its inception is estimated to be 6,000 hectares. Between 1985 and 1988, Jordan's Ministry of Agriculture, in collaboration with ACSAD, used contour terraces and ridges for pasture and range improvement in the Balama district. Better growth of olive, almond and pistachio was recorded on the experimental site.

In the Dei-Atiye community of Syria, rainwater harvesting was established in 1987 on an area of 130 ha. The project site was sub-divided into four parts for tree crops, range plants, cereals and runoff research. The International Centre for Agricultural Research in the Dry Areas (ICARDA) in Syria, is currently working on the improvement of various WH techniques and on the identification of waterharvesting areas suitable for various West Asian and North African (WANA) environments.

In the North-West Arabia, a local system known as "Mahafurs" is still in use. This system is simply a shallow excavation of 20 - 100 m in diameter surrounded on three sides by earthen bunds 1 - 4 m high. The open side is pointed in the direction of water flow inside the wadi bed and used to collect water for animal consumption and moisture for plant production. In Afghanistan, composite microcatchments have been in use for a long time. In a survey conducted in the early 1970s, over 70,000 ha of Meskat-type systems used for growing fruit trees were reported.

In tropical Asia, especially southern India and Sri Lanka, earth bunds and excavated hollows have been used for runoff retention during the rainy seasons for millenia. Tank storage permits farmers to grow a second dry season irrigated crop in addition to rainfed agriculture. Tanks are sited randomly so that any overland flow from one is caught by others downslope. In case of siltation, labourers are hired to remove the silts and spread them on the cropping land during the dry season.

In different parts of Libya, experimental sites of contour-ridge terracing covering more than 53,000 ha have recently been established. In 1990, the government of Tunisia started the implementation of the National Strategy of Surface Runoff Mobilisation which aims, among other things, at building 21 dams, 203 small earth dams, 1,000 ponds, 2,000 works to recharge water tables and 2,000 works for irrigation through water spreading by the year 2,000.

In Wadi El-Arish region of Egypt, stone dykes were used to direct the runoff water flow for irrigation purposes. Also cisterns, which store water meant for animal and human consumption as well as for supplemental irrigation, are common in Egypt. The number of cisterns has increased from less than 3,000 in 1960 to about 15,000 in 1993 with a capacity of about 4 million m^3. In the North-West region of Egypt a GTZ/FAO sponsored project on land use planning including water harvesting activities was carried out (El-Shafey).

In Yemen, small dams storing runoff for later use in irrigation or rural supply have been constructed since the beginning of the eighties; the total storage capacity is between 50,000 to 90,000 m^3. "Matfia" is an old technique of storing water for human and animal consumption in Morocco, which still continues. Expensive modern technology, including the use of reinforced concrete has now been introduced in constructing the cisterns, although the local people are less interested in these large and expensive systems.

Since 1984, Morocco has started constructing dams to harvest flood water. The upstream catchment area under these dams ranges from 500 to 10,000 hectares. As of 1988, thirty five of these dams had been constructed. They provide irrigation water for about 160,000 animals and 3,000 ha of cultivated plots.

In Ethiopia, the Sudan and Botswana, small check-dams made of earth are used to catch moderate overland flow passing down slight slopes. They are called "haffirs" and support crops planted upslope.

In 1988, special ploughs developed by Italian scientists were used in Niger for the implementation of microcatchments on a large scale. Results from these plots showed excellent rates of tree establishment. A set of test plots on improved trapezoidal bunds in Baringo, Kenya has been constructed. This improved version consists of earth bunds which surround the plot and diverge as collection arms upslope to increase the catchment area. In 1984, a self-help project sponsored by Oxfam and known as Turkana Water Harvesting Project, was started in the Turkana district of Kenya. It was aimed at developing systems of water harvesting for crop production, while also introducing animal ploughing.

In Western Australia, topography modification in the form of catchment treatment has been practised for a long time. These are known as "roaded" catchments. They consist of parallel ridges ("roads") of steep, bare and compacted earth, surveyed at a gradient that allows runoff to occur without causing erosion of the intervening channels. In 1980, it was estimated that there were more than 3,500 roaded catchment systems in Western Australia, and many of them have a top dressing or a layer of compacted clay to increase the runoff efficiency.

Numerous water harvesting projects have failed because the technology used proved to be unsuitable for the specific conditions of the site. Each of the water harvesting methods has its advantages and limitations which can be summarised as follows: (1) Water harvesting for animal consumption In developing countries, the building or reactivation of cisterns and other rainwater tanks for animal consumption can save water which othewise has to be lifted or pumped from groundwater or carried over long distances.

During recent years some technological developments took place in regard to water harvesting which might have some impact on the future role of WH in general:

i) *Supplemental water system*: Runoff water is collected and stored offside for later application to the cropped area using some irrigation method. The water stored allows a prolongation of the cropping season or a second crop.

ii) *Dual purpose systems*: In a dual purpose system the runoff water flows first through the crop area, then the excess water is stored in some facility for later irrigation use. In Arizona, USA, runoff irrigation was combined e. g. with trickle irrigation, using sealed soil surfaces to increase runoff rates.

iii) *Combined systems*: If the irrigation water from aquifers or from rivers/ reservoirs is not sufficient for year-round irrigation, a combination with runoff-irrigation (during the rainy season) is feasible. The combination of runoff-and furrow irrigation is reported from North Central Mexico.

iv) *Modelling*: If more information on hydrological, soil and crop parameters is available, models can be developed and applied to water harvesting for certain environments.

Rainwater Harvesting Systems

Typically, a rainwater harvesting system consists of three basic elements: the collection system, the conveyance system, and the storage system. Collection systems can vary from simple types within a household to bigger systems where a large catchment area contributes to an impounding reservoir from which water is either gravitated or pumped to water treatment plants.

The categorisation of rainwater harvesting systems depends on factors like the size and nature of the catchment areas and whether the systems are in urban or rural settings. Some of the systems are described below.

Simple Roof Water Collection Systems

While the collection of rainwater by a single household may not be significant, the impact ofthousands or even millions of household rainwater storage tanks can potentially be enormous. The main components in a simple roof water collection system are the cistern itself, the piping that leads to the cistern and the appurtenances within the cistern. The materials and the degree of sophistication of the whole system largely depend on the initial capital investment. Some cost effective systems involve cisterns made with ferro-cement, etc. In some cases, the harvested rainwater may be filtered. In other cases, the rainwater may be disinfected.

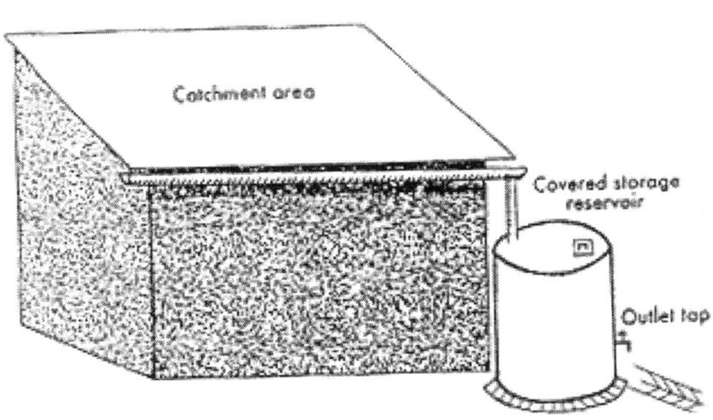

Figure 1. Example of a roof catchment system

Larger Systems for Institutions, Stadiums, Airports, and Other Facilities

When the systems are larger, the overall system can become a bit more complicated, for example rainwater collection from the roofs and grounds of institutions, storage in underground reservoirs, treatment and then use for non-potable applications.

Roof Water Collection Systems for High-rise Buildings

In high-rise buildings, roofs can be designed for catchment purposes and the collected roof water can be kept in separate cisterns on the roofs for non-potable uses.

Land surface catchments

Rainwater harvesting using ground or land surface catchment areas can be a simple way of collecting rainwater. Compared to rooftop catchment techniques, ground catchment techniques provide more opportunity for collecting water from a larger surface area. By retaining the flows (including flood flows) of small creeks and streams in small storage reservoirs (on surface or underground) created by low cost (e.g., earthen) dams, this technology can meet water demands during dry periods. There is a possibility of high rates of water loss due to infiltration into the ground, and because of the often marginal quality of the water collected, this technique is mainly suitable for storing water for agricultural purposes.

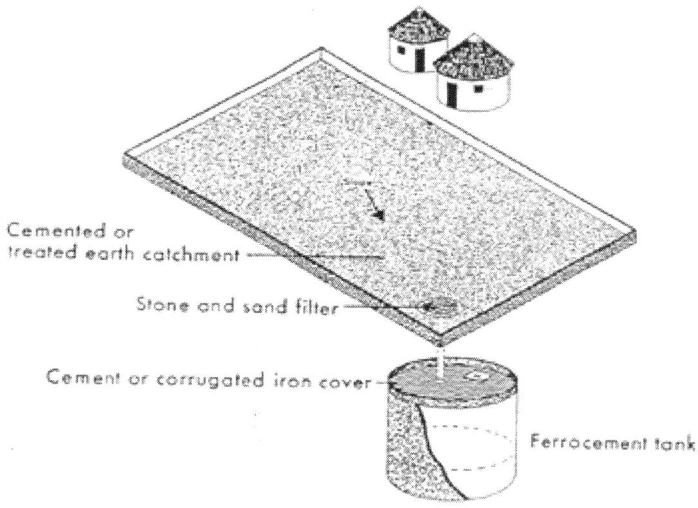

Figure 2. Example of a ground catchment system

Collection of Stormwater in Urbanised Catchments

The surface runoff collected in stormwater ponds/reservoirs from urban areas is subject to a wide variety of contaminants. Keeping these catchments clean is of primary importance, and hence the cost of water pollution control can be considerable.

Rainwater Conservation: Applications to Urban Areas

Many cities around the world obtain their water from great distances—often over 100km away. But this practice of increasing dependence on the upper streams of the water resource supply area is not sustainable. Building dams in the upper watershed often means submerging houses, fields and wooded areas. It can also cause significant socio-economic and cultural impacts in the affected communities. In addition, some existing dams have been gradually filling with silt. If not properly maintained by removing these sediments, the quantity of water collected may be significantly reduced.

When the city increases the degree of its dependence on a remote water resource, and there is a long period without rainfall in the upstream dam sites, the ability of the city to function effectively is seriously compromised. The same can be said about a city's reliance on a pipeline for drawing water from a water resource area to the city. A city which is totally reliant on a large, centralised water supply pipeline (or "life-line") is vulnerable in the face of a large-scale natural disaster. A shift from "life-line" to decentralised "life-points" should be encouraged. Numerous scattered water resource "life-points" within a city are more resilient and can draw on rainwater and groundwater, providing the city with greater flexibility in the face of water shortages and earthquakes.

Due to the rapid pace of urbanisation, many of the world's large cities are facing problems with urban floods. The natural hydrological cycle manifests itself at different scales, depending upon climatic, geographic and biological factors. As rain falls over time and seeps underground to become groundwater, it feeds submerged springs and rivers. The concrete and asphalt structures of cities have tended to disrupt the natural hydrological cycle, and reduce the amount of rainwater permeating underground. A decrease in the area where water can penetrate speeds up the surface flow of rainwater, causing water to accumulate in drains and streams within a short time.

Every time there is concentrated heavy rain, there is an overflow of water from drains, and small and medium sized rivers and streams repeatedly flood. These conditions can often lead to an outpouring of sewage into rivers and streams from sewer outlets and sewer pumping stations, thus contaminating the quality of urban streams and rivers.

Concrete and asphalt have a profound impact on the ecology of the city. These include:

i) *Drying of the city* – This happens as rivers and watercourses are covered, natural springs dry up, and greenery is cut down.

ii) *Heat pollution* – In some cities during the hot summer, an asphalt road at midday can reach temperatures of over 60°C. The heat expelled from air conditioners can further aggravate this.

In order to achieve a comprehensive solution to this problem, new approaches to urban development are required emphasising sustainability and the restoration of the urban hydrological cycle. Traditionally, storm sewer facilities have been developed based on the assumption that the amount of rainwater drained away will have to be increased. From the standpoint of preserving or restoring the natural water cycle, it is important to retain rainwater and to facilitate its permeation by preserving natural groundcover and greenery.

Concept of "Cycle Capacity"

"Cycle capacity" refers to the time that nature needs revive the hydrological cycle. The use of groundwater should be considered from the point of view of cycle capacity. Rain seeps underground and over time becomes shallow stratum groundwater. Then, over a very long period of time, it becomes deep stratum groundwater. For sustainable use of groundwater, it is necessary to consider the storage capacity for groundwater over time. If this is neglected and groundwater is extracted too quickly, it will disappear within a short time.

Demand Side Management of Water Supply

In establishing their water supply plans, cities have usually assumed that the future demand for water will continue to increase. Typically, city waterworks departments have made excessive estimates of the demand for water and have built waterworks infrastructure based on the assumption of continued development of water resources and strategies to enlarge the area of water supply.

The cost of development is usually recovered through water rates, and when there is plenty of water in the resource area, conservation of the resource is not promoted. This tends to create a conflict when drought occurs, due to the lack of policies and programmes to encourage water conservation. It has even been suggested that the lack of promotion of water conservation and rainwater harvesting is due to the need to recover infrastructure development costs through sales of piped water. The exaggerated projection of water demand leads to the over-development of water resources, which in turn encourages denser population and more consumption of water.

Sustainability of urban water supply requires a change from coping with water supply without controlling demand, to coping with supply by controlling demand. The introduction of demand side management encourages all citizens to adopt a water conservation approaches, including the use of freely available, locally supplied rainwater.

Health Aspects in Utilising Rainwater

In the past, it was believed that rainwater was pure and could be consumed without pre-treatment. While this may be true in some areas that are relatively unpolluted, rainwater collected in many locations contains impurities. Particularly during the last three decades, "acid rain" has affected the quality of the collected water, to the point where it now usually requires treatment. Rainwater quality varies for a number of reasons. While there are widely accepted standards for drinking water, the development of approved standards for water when it is used for non-potable applications would facilitate the use of rainwater sources.

In terms of physical-chemical parameters, collected roof water, rainwater and urban storm water tend to exhibit quality levels that are generally comparable to the World Health Organisation (WHO) guideline values for drinking water. However, low pH rainwater can occur as a result of sulphur dioxide, nitrous oxide and other industrial emissions, hence air quality standards must be reviewed and enforced. In addition, high lead values can sometimes be attributed to the composition of certain roofing materials – thus it is recommended that for roof water collection systems, the type of roofing material should be carefully considered. A number of collected rainwater samples have exceeded the WHO values in terms of total coliform and faecal coliform. The ratios of faecal coliform to faecal streptococci from these samples indicated that the source of pollution was the droppings of birds, rodents, etc.

Currently, water quality control in roof water collection systems is limited to diverting first flushes and occasional cleaning of cisterns. Boiling, despite its limitations, is the easiest and surest way to achieve disinfection, although there is often a reluctance to accept this practice as taste is affected.

Chlorine in the form of household bleach can be used for disinfection, however the cost of UV disinfection systems are usually prohibitive. One promising area of research is the use of photo-oxidation based on available sunlight to remove both the coliforms and streptococci.

Design and Maintain Facilities

Catchment Surface

The effective catchment area and the material used in constructing the catchment surface influence the collection efficiency and water quality. Materials commonly used for roof catchment are corrugated aluminium and galvanised iron, concrete, fibreglass shingles, tiles, slates, etc. Mud is used primarily in rural areas. Bamboo roofs are least suitable because of possible health hazards. The materials of catchment surfaces must be non-toxic and not contain substances which impair water quality.

For example, asbestos roofs should be avoided; also, painting or coating of catchment surfaces should be avoided if possible. If the use of paint or coating is unavoidable, only non-toxic paint or coating should be used; lead, chromium, and zinc-based paints/coatings should be avoided. Similarly, roofs with metallic paint or other coatings are not recommended as they may impart tastes or colour to the collected water. Catchment surfaces and collection devices should be cleaned regularly to remove dust, leaves and bird droppings so as to minimise bacterial contamination and maintain the quality of collected water. Roofs should also be free from over-hanging trees since birds and animals in the trees may defecate on the roof.

When land surfaces are used as catchment areas, various techniques are available to increase runoff capacity, including: i) clearing or altering vegetation cover, ii) increasing the land slope with artificial ground cover, and iii) reducing soil permeability by soil compaction. Specially constructed ground surfaces (concrete, paving stones, or some kind of liner) or paved runways can also be used to collect and convey rainwater to storage tanks or reservoirs. In the case of land surface catchments, care is required to avoid damage and contamination by people and animals. If required, these surfaces should be fenced to prevent the entry of people and animals. Large cracks in the paved catchment due to soil movement, earthquakes or exposure to the elements should be repaired immediately. Maintenance typically consists of the removal of dirt, leaves and other accumulated materials. Such cleaning should take place annually before the start of the major rainfall season.

Conveyance Systems

Conveyance systems are required to transfer the rainwater collected on catchment surfaces (e.g. rooftops) to the storage tanks. This is usually accomplished by making connections to one or more down-pipes connected to collection devices (e.g. rooftop gutters). The pipes used for conveying rainwater, wherever possible, should be made of plastic, PVC or other inert substance, as the pH of rainwater can be low (acidic) and may cause corrosion and mobilisation of metals in metal pipes.

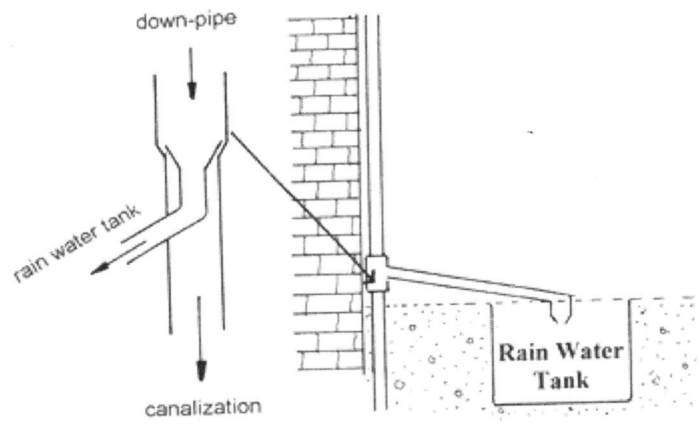

Figure 3. Example of a first flush device installation

When selecting a conveyance system, consideration should be given to the fact that when it first starts to rain, dirt and debris from catchment surfaces and collection devices will be washed into the conveyance systems (e.g. down-pipes). Relatively clean water will only be available sometime later in the storm. The first part of each rainfall should be diverted from the storage tank. There are several possible options for selectively collecting clean water for the storage tanks. The common method is a sediment trap, which uses a tipping bucket to prevent the entry of debris from the catchment surface into the tank. Installing a first flush (or foul flush) device is also useful to divert the initial batch of rainwater away from the tank.

Gutters and down-pipes need to be periodically inspected and carefully cleaned. A good time to inspect gutters and down-pipes is while it is raining, so that leaks can be easily detected. Regular cleaning is necessary to avoid contamination.

Storage Tanks

Storage tanks for collected rainwater may be located either above or below the ground. They may be constructed as a part of the building, or may be built as a separate unit located some distance away from the building. The design considerations vary according to the type of tank and other factors. Various types of rainwater storage facilities can be found in practice. Storage tanks should be constructed of inert material. Reinforced concrete, fibreglass, polyethylene, and stainless steel are suitable materials.

Ferro-cement tanks and jars made of mortar or earthen materials are commonly used. As an alternative, interconnected tanks made of pottery or polyethylene may be suitable. The polyethylene tanks are compact but have a large storage capacity (1,000 to 2,000 litres). They are easy to clean and have many openings which can be fitted with connecting pipes. Bamboo reinforced tanks are less successful because the bamboo may become infested with termites, bacteria and fungus.

Precautions are required to prevent the entry of contaminants into storage tanks. The main sources of external contamination are pollution from debris, bird and animal droppings, and insects that enter the tank. Sometimes, human, animal and other environmental contaminants, which happen to fall into tanks, can cause contamination. Open containers are not recommended for storing water for drinking purposes. A solid and secure cover is required to avoid breeding of mosquitoes, to prevent insects and rodents from entering the tank, and to keep out sunlight to prevent the growth of algae inside the tank. A coarse inlet filter is also desirable for excluding coarse debris, dirt, leaves, and other solid materials.

The storage tank should be checked and cleaned periodically. All tanks need cleaning and their designs should allow for thorough scrubbing of the inner walls and floors. A sloped bottom and the provision of a sump and a drain are useful for collection and discharge of settled grit and sediment. An entrance hole is required for easy access for cleaning. The use of a chlorine solution is recommended for cleaning, followed by thorough rinsing. Chlorination of the cisterns or storage tanks is necessary if the water is to be used for drinking and domestic uses. Dividing tanks into two sections or dual tanks can facilitate cleaning. Cracks in the storage tanks can create major problems and should be repaired immediately.

The extraction system (e.g., taps/faucets, pumps) must not contaminate the stored water. Taps/faucets should be installed at least 10 cm above the base of the tank as this allows any debris entering the tank to settle on the bottom, where if it remains undisturbed, will not affect the quality of the water. Rainwater pipes must be permanently marked in such a way that there is no risk of confusing them with drinking water pipes. Taps must also be clearly labelled for the user both in the local language and in clear graphic images. The handle of taps might be detachable to avoid the misuse by children. Periodic maintenance should also be carried out on any pumps used to lift water to selected areas in the house or building.

The following devices are also desirable.

i) An overflow pipe leading into either infiltration plants, drainage pipes with sufficient capacity or the municipal sewage pipe system.
ii) An indicator of the amount of water in the storage tank
iii) A vent for air circulation (often the overflow pipe can substitute)
iv) Protection against insects, rodents, vermin, etc. may also be required.

When using rainwater, it is important to recognise that the rainfall is not constant throughout the year; therefore, planning the storage system with an adequate capacity is required for the constant use of rainwater even during dry periods. Knowledge of the rainfall quantity and seasonality, the area of the catchment surface and volume of the storage tank, and quantity and period of use required for water supply purposes is critical. For example, in Tokyo, the average annual rainfall is about 1,400 mm. Assuming that the effective catchment area of a house is equal to the horizontal line of its roof surface area, and given that the roof surface area is 50 m^2, the average annual volume of rainwater falling on the roof may be calculated as 70 m^3. However, in practice, this volume can never be achieved since a portion of the rainwater evaporates from the roof surface and a portion may be lost to the drainage system, including the first flush. Furthermore, a portion of collected rainwater volume may be lost as overflow from the storage container if the storage tank has insufficient capacity to store the entire collected volume even in a heavy rain. Thus, the net usable or available amount of rainwater from the roof surface would be approximately 70% to 80% of the gross volume of rainfall. In the above example, the actual usable amount of rainwater would be about 49 m^3 to 56 m^3 in a year.

The disadvantages of the rainwater harvesting and utilisation systems are:

i) The catchment area and storage capacity of a system are relatively small. There is a great variation in weather. During a prolonged drought, the storage tank may dry up.

ii) Maintenance of rainwater harvesting systems, and the quality of collected water, can be difficult for users.

iii) Extensive development of rainwater harvesting systems may reduce the income of public water systems.

iv) Rainwater harvesting systems are often not part of the building code and lack clear guidelines for users/developers to follow.

v) Rainwater utilisation has not been recognised as an alternative of water supply system by the public sector. Governments typically do not include rainwater utilisation in their water management policies, and citizens do not demand rainwater utilisation in their communities.

vi) Rainwater storage tanks may be a hazard to children who play around it.

viii) Rainwater storage tanks may take up valuable space.

ix) Some development costs of larger rainwater catchment system may be too high if the costs are not shared with other systems as part of a multi-purpose network

Systemic Design of Rainwater Utilisation System

Rainwater utilisation, together with water conservation and wastewater reclamation, should be incorporated into municipal ordinances and regulations. Some standardisation of materials, at least at regional level, may be desirable from a maintenance and replacement point of view. It may be also appropriate to standardise the design of the rainwater utilisation system, at least at the regional level.

Implementation Policies

Various implementation policies should be established to make rainwater utilisation and other measures a part of the social system. Leadership is very important and local governments must take the initiative to promote the concept of water resource independence and restoration of the natural hydrological cycle. Consideration should be given to subsidising facilities for rainwater utilisation.

Technology and Human Resource Development

Encouraging technology and human resources development to support rainwater utilisation is very important. It is also important to promote the development of efficient and affordable devices to conserve water, facilities to use rainwater and devices to enhance the underground seepage of rainwater. Together with this, there is a need to train specialists with a thorough grasp of these technologies and devices.

Networking

To promote rainwater harvesting and utilisation as an environmentally sound approach for sustainable urban water management, a network should be established involving government administrators, citizens, architects, plumbers and representatives of equipment manufacturers. It is essential to encourage regional exchanges amongst public servants, citizens and industry representatives involved in rainwater storage, seepage and use, as well as the conservation and reclamation of water.

Public awareness and education are essential to improve acceptance and implementation of rainwater harvesting and utilisation.

Examples of Rainwater Harvesting and Utilisation

In Singapore, which has limited land resources and a rising demand for water, is on the lookout for alternative sources and innovative methods of harvesting water. Almost 86% of Singapore's population lives in high-rise buildings. A light roofing is placed on the roofs to act as catchment. Collected roof water is kept in separate cisterns on the roofs for non-potable uses. A recent study of an urban residential area of about 742 ha used a model to determine the optimal storage volume of the rooftop cisterns, taking into consideration non-potable water demand and actual rainfall at 15-minute intervals. This study demonstrated an effective saving of 4% of the water used, the volume of which did not have to be pumped from the ground floor. As a result of savings in terms of water, energy costs, and deferred capital, the cost of collected roof water was calculated to be S$0.96 against the previous cost of S$1.17 per cubic meter.

A marginally larger rainwater harvesting and utilisation system exists in the Changi Airport. Rainfall from the runways and the surrounding green areas is diverted to two impounding reservoirs. One of the reservoirs is designed to balance the flows during the coincident high runoffs and incoming tides, and the other reservoir is used to collect the runoff. The water is used primarily for non-potable functions such fire-fighting drills and toilet flushing. Such collected and treated water accounts for 28 to 33% of the total water used, resulting in savings of approximately S$ 390,000 per annum.

In Tokyo, rainwater harvesting and utilisation is promoted to mitigate water shortages, control floods, and secure water for emergencies. The Ryogoku Kokugikan Sumo-wrestling Arena, built in 1985 in Sumida City, is a well-known facility that utilises rainwater on a large scale. The 8,400 m^2 rooftop of this arena is the catchment surface of the rainwater utilisation system. Collected rainwater is drained into a 1,000 m^3 underground storage tank and used for toilet flushing and air conditioning. Sumida City Hall uses a similar system. Following the example of Kokugikan, many new public facilities have begun to introduce rainwater utilisation systems in Tokyo.

At the community level, a simple and unique rainwater utilisation facility, "Rojison", has been set up by local residents in the Mukojima district of Tokyo to utilise rainwater collected from the roofs of private houses for garden watering, fire-fighting and drinking water in emergencies. To date, about 750 private and public buildings in Tokyo have introduced rainwater collection and utilisation systems. Rainwater utilisation is now flourishing at both the public and private levels.

In October 1998, rainwater utilisation systems were introduced in Berlin as part of a large scale urban re-development, the DaimlerChrysler Potsdamer Platz, to control urban flooding, save city water and create a better micro climate. Rainwater falling on the rooftops (32,000 m^2) of 19 buildings is collected and stored in a 3500 m^3 rainwater basement tank. It is then used for toilet flushing, watering of green areas (including roofs with vegetative cover) and the replenishment of an artificial pond.

In another project at Belss-Luedecke-Strasse building estate in Berlin, rainwater from all roof areas (with an approximate area of 7,000 m^2) is discharged into a separate public rainwater sewer and transferred into a cistern with a capacity of 160 m^3, together with the runoff from streets, parking spaces and pathways (representing an area of 4,200 m^2). The water is treated in several stages and used for toilet flushing as well as for garden watering. The system design ensures that the majority of the pollutants in the initial flow are flushed out of the rainwater sewer into the sanitary sewer for proper treatment in a sewage plant. It is estimated that 58% of the rainwater can be retained locally through the use of this system. Based on a 10-year simulation, the savings of potable water through the utilisation of rainwater are estimated to be about 2,430 m^3 per year, thus preserving the groundwater reservoirs of Berlin by a similar estimated amount.

Both of these systems not only conserve city water, but also reduce the potential for pollutant discharges from sewerage systems into surface waters that might result from stormwater overflows. This approach to the control of non point sources of pollution is an important part of a broader strategy for the protection of surface water quality in urban areas.

Storing rainwater from rooftop run-off in jars is an appropriate and inexpensive means of obtaining high quality drinking water in Thailand. Prior to the introduction of jars for rainwater storage, many communities had no means of protecting drinking water from waste and mosquito infestation. The jars come in various capacities, from 100 to 3,000 litres and are equipped with lid, faucet, and drain. The most popular size is 2,000 litres, which costs 750 Baht, and holds sufficient rainwater for a six-person household during the dry season, lasting up to six months.

Two approaches are used for the acquisition of water jars. The first approach involves technical assistance and training villagers on water jar fabrication. This approach is suitable for many villages, and encourages the villagers to work cooperatively. Added benefits are that this environmentally appropriate technology is easy to learn, and villagers can fabricate water jars for sale at local markets. The second approach is applicable to those villages that do not have sufficient labour for making water jars. It involves access to a revolving loan fund to assist these villages in purchasing the jars. For both approaches, ownership and self-maintenance of the water jars are important. Villagers are also trained on how to ensure a safe supply of water and how to extend the life of the jars.

Initially implemented by the Population and Community Development Association (PDA) in Thailand, the demonstrated success of the rainwater jar project has encouraged the Thai government to embark on an extensive national program for rainwater harvesting.

In Indonesia, groundwater is becoming more scarce in large urban areas due to reduced water infiltration. The decrease of groundwater recharge in the cities is directly proportional to the increase in the pavement and roof area. In addition, high population density is has brought about high groundwater consumption. Recognising the need to alter the drainage system, the Indonesian government introduced a regulation requiring that all buildings have an infiltration well. The regulation applies to two-thirds of the territory, including the Special Province of Yogyakarta, the Capital Special Province of Jakarta, West Java and Central Java Province. It is estimated that if each house in Java and Madura had its own infiltration well, the water deficit of 53% by the year of 2000 would be reduced to 37%, which translates into a net savings of 16% through conservation.

In the Philippines, a rainwater harvesting programme was initiated in 1989 in Capiz Province with the assistance of the Canadian International Development Research Centre (IDRC). About 500 rainwater storage tanks were constructed made of wire-framed ferro-cement, with capacities varying from 2 to 10 m^3. The construction of the tanks involved building a frame of steel reinforcing bars (rebar) and wire mesh on a sturdy reinforced concrete foundation. The tanks were then plastered both inside and outside, thereby reducing their susceptibility to corrosion relative to metal storage tanks.

The rainwater harvesting programme in Capiz Province was implemented as part of an income generation initiative. Under this arrangement, loans were provided to fund the capital cost of the tanks and related agricultural operations. Loans of US$200, repayable over a three-year period, covered not only the cost of the tank but also one or more income generating activities such as the purchase and rearing of pigs, costing around US$25 each. Mature pigs can sell for up to US$90 each, providing an income opportunity for generating that could provide sufficient income to repay the loan. This type of innovative mechanism for financing rural water supplies can help avoid the requirement for water resources development subsidies.

In Bangladesh, rainwater collection is seen as a viable alternative for providing safe drinking water in arsenic affected areas. Since 1997, about 1000 rainwater harvesting systems have been installed in the country, primarily in rural areas, by the NGO Forum for Drinking Water Supply & Sanitation. This Forum is the national networking and service delivery agency for NGOs, community-based organisations and the private sector concerned with the implementation of water and sanitation programmes in unserved and underserved rural and urban communities. Its primary objective is to improve access to safe, sustainable, affordable water and sanitation services and facilities in Bangladesh.

The rainwater harvesting tanks in Bangladesh vary in capacity from 500 litres to 3,200 litres, costing from Tk. 3000-Tk.8000 (US$ 50 to US$ 150). The composition and structure of the tanks also vary, and include ferro-cement tanks, brick tanks, RCC ring tanks, and sub-surface tanks. The rainwater that is harvested is used for drinking and cooking and its acceptance as a safe, easy-to-use source of water is increasing amongst local users. Water quality testing has shown that water can be preserved for four to five months without bacterial contamination. The NGO Forum has also undertaken some recent initiatives in urban areas to promote rainwater harvesting as an alternative source of water for all household purposes.

Gansu is one of the driest provinces in China. The annual precipitation is about 300 mm, while potential evaporation amounts to 1500-2000 mm. Surface water and groundwater is limited, thus agriculture in the province relies on rainfall and people generally suffer from inadequate supplies of drinking water. Since the 1980s, research, demonstration and extension projects on rainwater harvesting have been carried out with very positive results.

In 1995/96, the "121" Rainwater Catchment Project implemented by the Gansu Provincial Government supported farmers by building one rainwater collection field, two water storage tanks and providing one piece of land to grow cash crops. This project has proven successful in supplying drinking water for 1.3 million people and developing irrigated land for a courtyard economy. As of 2000, a total of 2,183,000 rainwater tanks had been built with a total capacity of 73.1 million m^3 in Gansu Province, supplying drinking water for 1.97 million people and supplementary irrigation for 236,400 ha of land.

Rainwater harvesting has become an important option for Gansu Province to supply drinking water, develop rain-fed agriculture and improve the ecosystem in dry areas. Seventeen provinces in China have since adopted the rainwater utilisation technique, building 5.6 million tanks with a total capacity of 1.8 billion m^3, supplying drinking water for approximately 15 million people and supplemental irrigation for 1.2 million ha of land.

Although in some parts of Africa rapid expansion of rainwater catchment systems has occurred in recent years, progress has been slower than Southeast Asia. This is due in part to the lower rainfall and its seasonal nature, the smaller number and size of impervious roofs and the higher costs of constructing catchment systems in relation to typical household incomes.

The lack of availability of cement and clean graded river sand in some parts of Africa and a lack of sufficient water for construction in others, add to overall cost.

Nevertheless, rainwater collection is becoming more widespread in Africa with projects currently in Botswana, Togo, Mali, Malawi, South Africa, Namibia, Zimbabwe, Mozambique, Sierra Leone and Tanzania among others. Kenya is leading the way. Since the late 1970s, many projects have emerged in different parts of Kenya, each with their own designs and implementation strategies. These projects, in combination with the efforts of local builders called "fundis" operating privately and using their own indigenous designs, have been responsible for the construction of many tens of thousands of rainwater tanks throughout the country. Where cheap, abundant, locally available building materials and appropriate construction skills and experience are absent; ferro-cement tanks have been used for both surface and sub-surface catchment.

Due to inadequate piped water supplies, the University of Dar es Salaam, Tanzania has applied rainwater harvesting and utilisation technology to supplement the piped water supply in some of the newly built staff housing. Rainwater is collected from the hipped roof made with corrugated iron sheets and led into two "foul" tanks, each with a 70-litre capacity. After the first rain is flushed out, the foul tanks are filled up with rainwater. As the foul tanks fill up, settled water in the foul tanks flows to two underground storage tanks with a total capacity of 80,000 litres.

Then, the water is pumped to a distribution tank with 400 litres capacity that is connected to the plumbing system of the house. The principles for the operation of this system are: (i) only one underground tank should be filled at a time; (ii) while one tank is being filled, water can be consumed from the other tank, (iii) rainwater should not be mixed with tap water; (iv) underground storage tanks must be cleaned thoroughly when they are empty; (v) in order to conserve water, water should only be used from one distribution tank per day.

Thousands of roof catchment and tank systems have been constructed at a number of primary schools, health clinics and government houses throughout Botswana by the town and district councils under the Ministry of Local Government, Land and Housing (MLGLH). The original tanks were prefabricated galvanised steel tanks and brick tanks. The galvanised steel tanks have not performed well, with a short life of approximately 5 years.

The brick tanks are unpopular, due to leakage caused by cracks, and high installation costs. In the early 1980s, the MLGLH replaced these tanks in some areas with 10-20 m^3 ferro-cement tanks promoted by the Botswana Technology Centre. The experience with ferro-cement tanks in Botswana is mixed; some have performed very well, but some have leaked, possibly due to poor quality control.

Over the past decade, many NGOs and grassroots organisations have focused their work on the supply of drinking water using rainwater harvesting, and the irrigation of small-scale agriculture using sub-surface impoundments. In the semi-arid tropics of the north-eastern part of Brazil, annual rainfall varies widely from 200 to 1,000 mm, with an uneven regional and seasonal rainfall pattern. People have traditionally utilised rainwater collected in hand-dug rock catchments and river bedrock catchments.

To address the problem of unreliable rural drinking water supply in north-eastern Brazil, a group of NGOs combined their efforts with government to initiate a project involving the construction of one million rainwater tanks over a five year period, with benefits to 5 million people. Most of these tanks are made of pre-cast concrete plates or wire mesh concrete. Rainwater harvesting and utilisation is now an integrated part of educational programs for sustainable living in the semi-arid regions of Brazil. The rainwater utilisation concept is also spreading to other parts of Brazil, especially urban areas. A further example of the growing interest in rainwater harvesting and utilisation is the establishment of the Brazilian Rainwater Catchment Systems Association, which was founded in 1999 and held its 3rd Brazilian Rainwater Utilisation Symposium in the fall of 2001.

The island of Bermuda is located 917 km east of the North American coast. The island is 30 km long, with a width ranging from 1.5 to 3 km. The total area is 53.1 km^2. The elevation of most of the land mass is less than 30 m above sea level, rising to a maximum of less than 100 m. The average annual rainfall is 1,470 mm. A unique feature of Bermuda roofs is the wedge-shaped limestone "glides" which have been laid to form sloping gutters, diverting rainwater into vertical leaders and then into storage tanks.

Most systems use rainwater storage tanks under buildings with electric pumps to supply piped indoor water. Storage tanks have reinforced concrete floors and roofs, and the walls are constructed of mortar-filled concrete blocks with an interior mortar application approximately 1.5 cm thick. Rainwater utilisation systems in Bermuda are regulated by a Public Health

Act which requires that catchments be whitewashed by white latex paint; the paint must be free from metals that might leach into water supplies. Owners must also keep catchments, tanks, gutters, pipes, vents, and screens in good repair. Roofs are commonly repainted every two to three years and storage tanks must be cleaned at least once every six years.

St. Thomas, US Virgin Islands, is an island city which is 4.8 km wide and 19 km long. It is situated adjacent to a ridge of mountains which rise to 457 m above sea level. Annual rainfall is in the range of 1,020 to 1,520 mm. A rainwater utilisation system is a mandatory requirement for a residential building permit in St. Thomas. A single-family house must have a catchment area of 112 m^2 and a storage tank with 45 m^3 capacity. There are no restrictions on the types of rooftop and water collection system construction materials.

Many of the homes on St. Thomas are constructed so that at least part of the roof collects rainwater and transports it to storage tanks located within or below the house. Water quality test of samples collected from the rainwater utilisation systems in St. Thomas found that contamination from faecal coliform and Hg concentration was higher than EPA water quality standards, which limits the use of this water to non-potable applications unless adequate treatment is provided.

At the U.S. National Volcano Park, on the Island of Hawaii, rainwater utilisation systems have been built to supply water for 1,000 workers and residents of the park and 10,000 visitors per day. The Park's rainwater utilisation system includes the rooftop of a building with an area of 0.4 hectares, a ground catchment area of more than two hectares, storage tanks with two reinforced concrete water tanks with 3,800 m^3 capacity each, and 18 redwood water tanks with 95 m^3 capacity each. Several smaller buildings have their own rainwater utilisation systems as well. A water treatment and pumping plant was built to provide users with good quality water.

Water Harvesting Systems for Agriculture

Water harvesting is a technique of developing surface water resources that can be used in dry regions to provide water for livestock, for domestic use, and for agroforestry and small scale subsistence farming.

Water harvesting systems may be defined as artificial methods whereby precipitation can be collected and stored until it is beneficially used. The system includes: 1) a catchment area, usually prepared in some manner to improve run off efficiency and 2) a storage facility for the harvested water, unless the water is to be immediately concentrated in the soil profile of a smaller area for growing drought-hardy plants. A water distribution scheme is also required for the systems devoted to subsistence farming for irrigation during dry periods.

The earliest evidence of the use of water harvesting are the well publicized systems used by the people of the Negev Desert perhaps 4000 years ago. Hillsides were cleared of vegetation and smoothed in order to provide as much run off as possible; the water was then channelled in contour ditches to agricultural fields and/or to cisterns. By the time the Roman Empire extended into the region, this method of farming encompassed more than 250,000 hectares and had become quite sophisticated.

In the New World, about 400-700 years ago, people living in North America in what is now the state of Colorado in the United States, and those living in what is now Peru in South America employed relatively simple methods of water harvesting for irrigation.

Although the practice of collecting water from rooftops is a very ancient type of water harvesting, practiced since the earliest of times up to the present, the harvesting of rain water for agriculture and ranching was pioneered in Australia during the 1920's. Galvanized rooflike structures were built near the ground surface solely for the purpose of water harvesting.

A renewed interest in the technology of water harvesting occurred in the 1950's in Israel, Australia and the United States. In Australia, 'roaded' catchments based on the concept of compacted earth were constructed over more than 2,000 hectares in order to collect water for agricultural purposes. In the United States, at about the same time, catchments were constructed primarily of sheet metal for watering livestock. Also during this period, experimentation was undertaken with plastic and artificial rubber membranes for the construction of both catchments and reservoirs. Since the 1950's the technology of water harvesting has developed variety and sophistication through experimentation and demonstration projects. Not all have been successful but some have worked very well.

The technology and experience gained has now reached the point where some form of water harvesting system can be designed to fit within the physical, climatological and economic constraints of almost any dry region of the world.

The development of water resources in arid lands at present and in the recent past has been concentrated on large-scale irrigation projects. Some have water supplied by river systems such as the Indus, Nile, Yellow River and Tigrus and Euphrates. Others depend upon ancient and finite groundwater bodies. However, the availability of modern technology to societies unprepared for it, combined with the common political focus on urgent and short-term goals, has created long-term and irrevocable consequences in many countries.

Presently, primarily because of salinization, as much land is being permanently retired from agriculture along major river systems as there is new land being added. Electrical and engine-driven pumps and improved methods of well construction have greatly increased the capacity of wells. But in many arid lands the absence of effective controls on drilling and pumping has resulted in groundwater mining with the consequences of decreasing yields, deteriorating water quality, saline intrusion and wastage of water.

Large irrigation projects are essential to many national economies in arid lands. They can be effective over the long term if there is not only a technical competence but an informed social and political structure as well. However, these projects provide few direct benefits to the small land holder or nomad who must exist within the constraints of his environment without the benefit of new technology appropriate to his needs.

Water harvesting offers one method of improving the livelihood of these people. It is not a panacea, but depending upon the rainfall regime in a specific area, it can be used to augment existing supplies, improve range utilization and provide food for better nutrition, wood for fuel and, perhaps more importantly, help towards economic stability by reducing the uncertainty of human life in arid ecosystems.

Much of the economy of arid lands depends upon livestock; so it is not surprising that most of the work that has been accomplished in water harvesting has been aimed at providing water for livestock. Many of the systems devised have been very effective.

Many rangeland areas are overgrazed around water sources, whereas large portions of the land and forage are unused because of inadequate water.

A uniform distribution of watering spots could allow these lands to be used for maximum benefit. Water harvesting can provide an adequate distribution of watering spots in most situations, and, with water, even dead forage can be utilized. However, there is an inherent danger that, in the absence of grazing controls, each new watering source can lead to increased herd size, overgrazing and a new nucleus of expanding desert. Just as with other forms of water resource development, in order to be lastingly effective, water harvesting projects and particularly those designed for livestock production must be preceeded by comprehensive planning and controls with provisions for perpetual operational support.

The purpose of water harvesting is to either augment existing water supplies or to provide water where other sources are either not available or would entail prohibitive developmental costs. The aim is to provide this water in sufficient quantity and of a suitable quality for the intended use.

Arid zones are often described as pulse-reserve systems that turn on with water, store material for the dry interval and then shut down until the next water event. The natural ecosystem has developed under these constraints but the strain on human populations is considerable, particularly on sedentary populations. The idea of water harvesting is to smooth out the peaks between want and abundance by collecting and storing water for the period of want, and in addition, to work within the arid ecosystem using a renewable resource.

Configuration and Use

The geometric configuration of water harvesting systems depends upon the topography, the type of catchment treatment, the intended use and personal preference. Microcatchments, strip harvesting, roaded catchments and harvesting aprons are some of the more common types.

Microcatchments and strip harvesting can be successful in years of normal or above normal rainfall. However, in dry years most annual crops will fail. They are best suited for agroforestry where drought resistant trees or other drought hardy perennial species are grown. The use of microcatchments involves the preparation of small catchment areas in which one to several plants are grown on the low side. The collection area may range from 20 to 1000 m depending upon the precipitation in the area and plant requirements. The microcatchment procedure may be used in complex terrain or on steep slopes where other water-harvesting techniques may be difficult to install.Strip farming is a modification of the microcatchment method. Berms are erected on the contour and the area between them prepared to serve as the collection area. Run off occurring between the berms is then concentrated above the downslope berm to irrigate the vegetation planted there. Only very drought hardy plants should be grown with this type of system.

Apron type water harvesting systems are used primarily for livestock, wildlife and domestic water supplies. The catchment area (apron) is treated to obtain a high run off efficienty, unless an existing impermeable surface is used. Gravel covered, asphalt impregnated fiberglass is a common treatment in the United States. The systems are designed for minimum maintenance and must be fenced. A storage tank with evaporation control is required with the necessary pipes and valves to conduct the water to drinking troughs or to households. The apron type system is the simplest to design. As a first approximation for the size of apron required, the following equation is helpful:

$$A = 1.13\ U/\ p \text{ where:}$$

A = catchment area m^2

U = annual water requirement litres

P = average annual precipitation mm

Roaded catchments are well suited to agroforestry and for growing high value horticultural crops (e.g. fruit and nut trees, grapes etc). They are also suitable for providing livestock water. These catchments are best adapted to very gently sloping ground.

A roaded catchment consists of parallel rows of drainages 100 m or less long and spaced 15 to 18 meters apart. Trees or horticultural species are planted in the drainages. Grapes have proven an excellent crop in Arizona, U.S.

The areas between drainages are shaped much like high-crowned roads to serve as catchments. Side slopes of the catchment roads and longitudinal slopes of the drainages should be no more than about 2% to prevent erosion. The catchments are cleared of vegetation and smoothed and are treated to reduce infiltration. Sodium chloride has been an effective treatment in Arizona. If high value, horticultural crops are to be grown, water storage is necessary to provide supplemental irrigation water. This is easily accomplished by diverting excess water from the drainages into a storage facility.

Water harvesting for agriculture requires a more complex system than do the other systems. The size of the catchment area in relation to that of the agricultural area must be balanced against crop demands topography, provided there is a level area to farm and care is taken with catchment construction to provide low slopes, or, in steep terrain, short slope lenghts broken by diversions.

Compartmental reservoirs consisting of three storages is recommended. Electric, engine or wind driven pumps are usually necessary to transfer water between ponds and to drive the irrigation system. In some steep terrains, gravity systems might be possible. Since fairly large quantities of water are usually required, efficient catchments are also necessary. But treatments can be expensive. Sodium chloride is one of the least expensive treatments in some areas and is effective in locations where the soil has a sufficient quantity (about 10% or more) of expanding clays.

Water harvesting for subsistence farming holds much promise in alleviating the food and nutrition problems of arid lands. Successful systems have been installed and shown great promise. However, training in new techniques is necessary, more information is needed from around the world on crop phenology and water requirements, and additional demonstration projects need to be installed under varying economic, social and climatological conditions in order to develop universal prescriptions.

Catchment Areas

The catchment area of any water harvesting system is an area that is reasonably impermeable to water which can be used to produce run off. Some examples are: l) natural surfaces such as rock outcrops, 2) surfaces developed for other purposes such as paved highways, aircraft runways, or rooftops, 3) surfaces prepared with minimal cost and effort such as those cleared of vegetation or rocks and smoothed, or both smoothed and compacted, 4) surfaces treated chemically with sodium salts, silicones latex or oils, 5) surfaces covered with asphalt, concrete, butyl rubber, metal foi, plastic, tarpaper or sheet metal.

The particular surface treatment selected will depend upon the availability of materials and labor and budget allotments. Generally, the greater the run off efficiency and life of treatment the greater the cost. At one end of the scale, simple smoothing and compact on of non porus soils is effective but requires annual maintenance. On the other end of the scale is asphalt impregnated fiberglass covered with gravel which may last 20 years or more.

Myers lists some desirable characteristics of catchment treatments as:

1. Run off from the surface must be non-toxic to man and animals.
2. The surface should be smooth and impermeable to water.
3. The treatment should have high resistance to weathering and should not deteriorate because of chemical or physical process.
4. The treatment need not have great mechanical strength but should be able to resist damage by hail or intense rainfall, wind, occasional animal traffic, moderate flow of water, plant growth, insects, birds and burrowing animals.
5. The material should be inexpensive on an annual cost basis and should require minimum site preparation.
6. Maintenance should be simple.

Obviously, there is no single treatment that would have all of these characteristics. Some trade off is necessary, but lowest cost over the long term is often the overgrazing objective.

Selection of the lowest cost catchment can sometimes be a mistake. For example, simple smoothed catchments produce water at very low cost, but they do not provide run off from small storms that characterize rainfall periods in many arid zones. A large and expensive structure might have to be built to store water for use during the period when no run off occurs. The cost of a simple smoothed catchment, plus the large storage required could be greater than the cost of a more expensive catchment which provides run off from small storms. A procedure has been developed that can be used for livestock and domestic water systems to determine the lowest cost for any given combination of unit construction costs of catchment and storage, catchment efficiency, water demand schedules and precipitation patterns.

Water Storage

There are basically three types of storages: 1) the soil profile, 2) excavated ponds and 3) tank or cistern containers. The early, ancient water harvesting systems were simple arrangements where water was directed from hillsides onto cultivated areas with the idea of immediately storing the water in the soil for plant use. The problem was whether sufficient water could be stored to offset a prolonged drought. However, the method is still valid today and can be used in agroforestry to grow drought resistant varieties of trees and other economic plants.

Excavated ponds are often the only economical means of storing the large quantities of water which are needed for run off farming. But evaporation and seepage are serious problems. Evaporation suppression on water impoundments is still in the experimental stage. Surface area reduction, reflective methods, surface films, mechanical covers, floating styrofoam balls, and even empty, plastic film canisters have all been used. Most have been effective to some extent, but a simple economical method has yet to be developed. Surface films are generally not economical on small impoundments. Reflective films are generally not economical on small impoundments. Reflective methods (beads or dyes floating on the surface) are ineffective in windy conditions. Some types of floating mechanical covers and floats have worked well in experimental situations, but most are short-lived and all are expensive.

Two inexpenisve methods for reducing evaporative surface area that have been effective in some situations are the compartmented reservoir system and sand or rock filled reservoirs. The compartmented reservoir is a system of pumping water between two or three ponds so as to minimize total surface area. Sand or rock reservoirs are structures either deliberately filled with rock or designed to capture gravel and sand alluvium, as well as water. The dam is sometimes built in stages so that only coarse sediments are deposited. However, about 50% of the capacity will be lost after filling. A well is sunk behind or through the dam to draw out the stored water. After the water level has sunk to about one meter below the surface of the fill, evaporation effectively ceases.

In a water balance study of three ponds in southern Arizona, seepage acounted for 58 to 87% of the total annual water loss. Seepage control is simpler and less expensive than evaporation control. Chemical dispersing agents, bentonite, soil cement, membrane liners, asphalt, salt and simple compaction have all been used successfully to seal impoundments. Treatment method and application rate is determined by soil type, purpose of the impoundment, severity of wetting and drying cycles, and economics.

Tanks or cisterns can be used effectively for livestock watering and domestic supplies. Seepage and evaporation suppression are less difficult and less expensive. Any container capable of holding water is a potential water storage facility. External water storages are a necessary component for drinking water supply systems, and may also be a part of a run off farming system where the water is applied to the cropped area by some form of irrigation system. In many water-harvesting systems, the storage and water distribution facility is the most expensive single item, and may represent up to 50% of the total cost.

There is an almost infinite number of types, shapes, and sizes of wooden and reinforced plastic storages. Costs and availability are primary factors for determining the potential suitability of these storages. One common type of storage is a steel tank with vertical walls and with a concrete or other type of impermeable bottom. Storages constructed from concrete and plaster are relatively inexpensive, but require considerable hand labor. Roofs over the storages are a common technique of suppressing evaporation although they are usually expensive. Floating covers of low density synthetic foam rubber are an effective means of controlling evaporation from vertical walled, open topped storages, and are not expensive.

Constraints and Strategies

The particular strategy to be taken in developing a water harvesting system depends upon a number of constraints. Some of the more important that must be considered include:

1. The need for acceptance by the local community whether the system is to be used for livestock, domestic purposes, agroforestry, or farming.
2. The quantity and quality of water required to meet the need.
3. The availability of alternative, less expensive sources that could be developed.
4. The amount, seasonal distribution and variability of rainfall.
5. The materials, labor and machinery available, suitable for installing a water harvesting system are within budgetary limitations.
6. The provisions for maintenance.

Need and Acceptance

The need for water resource development in almost all arid lands is patent.

The problem is to reconcile this need with what may be accomplished. Furthermore, the user must be fully aware of the potential benefits as well as the limitations of a proposed system. Rural people in arid lands are understandably conservative. They cannot take chances with their survival on unproven methods. But the ultimate success of a water harvesting project depends on the full support of the user for proper operation and maintenance. The user must believe that the system is the best for his needs.

In areas where the concepts of water harvesting and run off farming are not fully accepted, the first system installed must be constructed from materials which require minimum maintenance and have maximum effectiveness. m e extra cost encountered in building a substantial system may be necessary to insure acceptance of the concept by the user. Once the concepts are accepted, it is often possible to utilize lower cost materials and techniques on subsequent units even though these lower cost systems may have a greater chance of failure or require additional effort from the user. If the user has been shown the ideas are valid, he is more likely to expend the extra effort to operate and maintain the system properly.

Water Quantity

With the exception of water harvesting systems that do not require water storage facilities (e.g. microcatchments and strip catchments), other systems must be designed to supply the quantity of water needed at the times it is needed.

For livestock the need depends upon the time of the year the system is expected to supply water, and this is determined by the grazing systems employed and the monthly distribution of rainfall. For most installations, many combinations of catchment and storage sizes will provide the desired quantities of water. The problem is to find the most economical combination. A method of doing this can be found in Frasier and Myers.

For domestic supplies, people require from 20 to 40 liters per day for cooking, drinking and washing, depending upon how they conserve water. me system should be designed to account for this minimum requirement in addition to any losses that would occur by evaporation or seepage from storage.

Water harvesting systems designed for agriculture are more difficult to design. There is little information on the minimum total water requirements of agricultural crops although consumptive use data abound for crops abundantly supplied with water. Of equal importance to the total water requirement is the timing of the water needs. Erie, French, Bucks and Harris and Doorenbos and Kassam provide the most useful information to date on consumptive use. The seasonal pattern of water use from initial establishment to harvest must be satisfied by the design of the water harvesting system. This type of information has been developed for many crops under extensive irrigation practices and is probably higher than needed for many run off farming applications. Relationships of this nature must be developed or estimated for proper design of agricultural water harvesting systems and matched with the water supply to determine frequency and amount of irrigation.

Water Quality

Water collected from a catchment can contain organisms and water-soluble impurities from windblown dust deposited on the surface or chemical pollutants directly from the treatment (e.g. salt, silicone, tars, oils, etc) or from weathering by-products created by deterioration of the treatment materials. Asphalt and certain plastics are deteriorated by sunlight and heat into water soluble products. Animals feces can be a source of bacteria and virus contamination if the area is not protected against entrance. However, the quality of water from most surface treatments is usually adequate for livestock, but filters are needed in most cases if the water is for human consumption. None of the surface treatments (even sodium treatment) that have been successful appear to affect plant production.

Alternative Sources of Water

Although water harvesting is not necessarily an expensive means of developing water, there may be other sources of water at or near a particular site that can be developed more cheaply or that would insure more reliable supplies. For instance, undeveloped or underdeveloped springs, a shallow groundwater table which may receive reliable recharge along a mountain front, perched water that might be developed with horizontal wells may offer possibilities. These types of sources should be thoroughly investigated before embarking on a project. All potential water sources should be evaluated with respect to number, location, yield, dependability, and quality. If there are other convenient sources that can be developed economically but are deficient in yield or dependability they may be used to supplement a water harvesting system. If water quality if poor (high salt content, for example) harvested water might provide sufficient dilution for the purpose of the need.

Incorporating intermittent water sources, such as ephemeral flows in waddles, into the total water supply system can in some cases permit the installation of a smaller water harvesting facility. The harvested water can be saved for periods when the ephemeral sources are insufficient or dry up entirely. This type of combination not only saves time and money but can be of major importance during extended drought periods.

Precipitation

The amount, timing and variability of rain which occurs during a season or year are the key factors that must be evaluated in designing a water harvesting system. Long term daily records are the most desirable. In arid lands, at least 15 to 20 years of record are needed. If there are large variations between years, data from the two wettest years should be eliminated. If sufficient long-term data are available, stochastic methods can be used to determine the probabilities of extreme periods. Mean annual rainfall is not a very good indicator of available water because there will be more years with rainfall less than the mean than there will be years with rainfall greater than the mean.

To compensate for dry years, the size and efficiency of the catchment areas and storage can be increased. But regardless of the design, there will be risk involved because of the uncertainty of rainfall. The user must decide the amount of risk that can be accepted should there be insufficient rainfall during some periods.

In very general terms, if the annual rainfall is no more than about 150 mm, agroforestry, with plantings of indigenous or proven drought resistant species, using microcatchments or contour strips is suitable. Systems designed for livestock watering can also be used, but will require storage tanks protected from evaporation losses as well as efficient catchment surfaces. In areas with annual rainfall greater than 250 mm and if there are good water storage facilities, farming systems are possible. However, between 250 and 300 mm mean annual rainfall (unless the rainfall period coincides with the growing season) drought resistant crops should be used. At 300 mm and above almost any conventional crop can be grown. In locations with an average annual rainfall less than 50 to 80 mm, water harvesting will probably never be economically feasible.

The final size of the catchment area should be determined by computing a weekly or monthly water budget of collected water versus water requirement to help insure that there are no critical periods when there will be insufficient water.

Smaller systems can frequently be used when the periods of maximum rainfall coincide with periods of maximum use. Larger systems, with large storage capacity are necessary when the periods of greatest precipitation occur after the periods of greatest water needs when it may be necessary to store water for 6 to 9 months.

Materials and Labor

There is no universally best material for catchment and storage. The cost of alternative water sources and the importance of the water supply determine the costs which can be justified in a system. Ordinarily, the lowest cost of locally available materials are used. Usually, water-harvesting systems for supplying drinking water are constructed from materials which are more costly than can be economically justified for run off-farming applications. One must balance the cost of materials to the cost of labor. Some materials and installation techniques are labor intensive, but have a relatively low capital cost.

Maintenance

Failure to provide for maintenance will result in early failure of the system. Failure to repair minor damage can result in complete destruction of the system. A maintenance programme must be followed even when the water collected is not being used. Some types of catchment treatments and storages require more frequent and intense maintenance than others. However, most water harvesting systems can be adequately maintained with once a year inspection and repair visits if any problems that appear at other times are immediately repaired. All elements of the system should be inspected, including checks for leaks in any valves, pipes or water storages, as well as the condition of the catchment, weed, animal and insect control. Inspections and repair usually require only a few hours, but are as essential to the system as the initial installation.

Case Examples

Numerous water harvesting systems have already been installed in various arid lands of the world, and many more are being implemented. Although much *of* the work is in the experimental stages and it is perhaps too soon to evaluate the ultimate results. Most of the installations have reported success. Only a few representative cases are cited here.

Mexico

There are a number of water harvesting projects in Mexico varying from village water supply systems to run off farming. Two are given as examples.

Nuevo Leon

This system is composed of a set of 248 microcatchments of 70 m each for growing pistachio trees. Each tree is planted in a microcatchment. A major objective is to evaluate different catchment treatments. They are 1) compacted soil, 2) soda ash (Na_2CO_3), 3) road oil, 4) gravel covered polyethylene, 5) gravel covered asphalt and 6) smoothed soil. Soil moisture was monitored under each tree at depths of 15, 35, and 55 cm. Also included were tests of various soil coverings immediately around the trees to reduce evaporation.

This is a long term experimental demonstration project because of the slow growth of the trees. From preliminary observation it appears that soil eroded from the salt treated catchment is deposited around the tree. The deposited soil reduced infiltration before it could infiltrate into the soil.

Techo cuenca

A water harvesting system was constructed in 1975 to augment the domestic water supplies for 30 families (about 180 people) for the village of Laguntia y Ranchos Nuevos in the state of Nuevo Leon. The system consists of an inverted galvinized metal roof (269m) supported on a steel framework above an 80,000 liter steel tank. Village labor for constructing the system was 36% of the total cost of 143,000 1975 pesas. The system provides drinking water to the entire village for 4 and 1/2 months of the year, based on an allotment of 20 liters per day per family at about one third the cost that is incurred in hauling water.

India

Trials with okra were carried out to study the effects of different water conservation methods on plant growth and yield. It was found that out of 8 methods tried, the highest yield was obtained with plants in 2 rows 45 cm apart in furrows running east/west with ridges 60 cm apart. The beneficial effect of this method was attributed to the rainwater drainage from the ridges to the furrows.

Iran

Asphalt coverings were applied aw 1 liter/m^2 to slopes above the experimental run off plots and terraces (2 m wide) on a hillside near Tehran, Iran for harvesting rainwater for growing trees. Rainfall run off over a 5-year period was substantial. Robinia pseudoacacia, Cupressus arizonica and Fraxinus rotundifolia were planted. The data showed that there was a significant increase in height, diameter of stem and crown development of plants in asphalt treated plots as compared to controls.

Pakistan

The Pakistan Forest Institute installed microcatchments on more than 40 hectares in 1982 in an area that receives 250 to 300 mm of annual rainfall. The purpose of the project was to establish forest plantations. Species of Acacia, Prosopis, Tecoma and Parkinsonia were planted. Survival of species planted as control, without microcatchments, was only about 10%. Survival of species planted in microcatchments was 80 to 90%. Acacia tortilis had the highest survival rate and grew more than 4 feet per year.

Australia

Australia was among the first of the western countries to install operational water harvesting systems. The systems were designed to provide water for livestock and domestic needs. Much of this work was done in southwest Western Australia.

Harvesting from natural surfaces

Advantage has been taken of natural outcrops of granite which commonly occur on the crests of broad, sand plain ridges in the area. Run off from the 250 mm of annual rainfall on many of these outcrops is collected by means of low concrete or masonry gutters built around the lower slopes of the rock. The water so collected is conducted along the gutter to a concrete storage tank built on or close to the rock, or to an earth tank sited downslope from the rock. The water from these rock catchments is noted for its excellent quality. The Public Works Department has constructed harvesting systems of this sort on some of the larger outcrops to provide public supplies of potable water.

The Public Works Department of Western Australia initiated in 1948 a programme of construction of Roaded Catchments. These catchments consisted of clearing, shaping, and contouring to control length and degree of slope and compacting with the aid of pneumatic rollers. An estimated 2500 roaded catchments have been installed principally to supply water for livestock use. These average approximately two acres in size. There are also 21 roaded catchments totaling 1745 acres (706 hectares) and ranging in size from 30 to 175 acres (12.1 to 70.8 hectares) presently being used to furnish domestic water for small towns in Western Australia.

United States

The Indian reservation of southwestern United States have much in common with the developing countries of arid regions: remoteness, grazing cultures, lack of water and poor economies. A number of water harvesting systems have been tested by various agencies on these reservations. Three are cited here.

The Arizona Strip

The Arizona Strip lies across the Colorado River from the Hualpai Indian Reservation and south of the Arizona-Utah state line. The land is under the jurisdiction of the U.S. Department of the Interior, Bureau of Land Management, and is leased to cattlemen" for grazing by livestock.

Perennial streams and springs are rare, and groundwater is inaccessible due to depth and isolation of perched aquifers. Earthen reservoirs are often used, but are rarely dependable because of high seepage, evaporation losses, and low run off.

Two water harvesting systems were installed in September, 1974 to evaluate the potential of this technique for supplying the necessary animal drinking water. The catchment areas (0.3 and 0.4 hectares) were treated with a refined paraffin wax sprayed on the prepared soil surface. The collected water was stored in a 300,000 liter steel rim, concrete bottom tank with a foam rubber floating cover for controlling evaporation. Total cost was $ 8,925 and $ 9,150, including labor and miscellaneous items such as fencing and drinking troughs. The systems are maintained by the Bureau of Land Management.

During a drought in 1976-1977, these systems provided the only water supply. All other water sources went dry. Without this water, the permittees would have had to move their livestock. Ranchers have remarked that these systems were as good as, or better than, a spring. Since that time, the Bureau of Land Management has installed over 60 more units of various types and treatments, and several local ranchers are installing units on their own. There have been some failures of the later units, but this has not deterred the ranchers from accepting this method of water supply. This attitude has been developed because it was demonstrated that, with proper installation and maintenance, water harvesting can be effective method of water supply.

Black Mesa

The Black Mesa water-harvesting facility is located on the Navajo Indian Reservation in northeastern Arizona on displaced overburden from a strip coal mine. This is one of the largest systems in the United States. It consists of (1) three water storage ponds with a total capacity of slightly over 3 million litres, (2) two leveled agricultural terraces of 1 hectare each, (3) a "road" catchment for an orchard of 0.5 hectares, 84) a fiberglass asphalt-gravel catchment of 3.2 hectares, and a 2.9 hectare salt-treated catchment. A pump system is used to transfer the collected water between ponds and to lift the water to irrigate the crop areas. Initially, flood irrigation was used, but later, a prinkler system was installed.

Annual crops grown and evaluated were beets, onions, turnips, potatoes, chard, lettuce, cabbage, tomatoes, squash, beans, pumpkins, melons, manger and corn. All crops, except tomatoes, did well, with some producing at levels above the national average. The value of the corn produced was the lowest of all crops. This was not unexpected, but corn is a traditional food in the area, and was planted for social reasons. Fruit trees had never been grown in the area before. All trees were growing well after 3 years, but it was too soon to determine the potential production of the varieties planted. The water-harvesting project yielded about $ 1,700 net revenues per cultivated hectare. Revenues are expected to increase when the orchards reach maturity.

Shungopovi

The village of Shungopovi is located upon Second Mesa, on the Hopi Indian Reservation in northern Arizona. The village, built on top of a sandstone rock mesa, had no source of water. From the time of first establishment, the villagers had carried water up from the valley, initially on foot and, later on, on the backs of burros. In the early 1930's, a small water-harvesting system was installed to partially relieve the water shortage of the village.

An area of approximately 1/3 hectare was set aside, cleared, and the loose soil removed to expose the sandstone bedrock. Below the area, a deep cistern was hewed into the rock, and a concrete roof constructed. This system was a functional part of the village water supply for about 30 years, at which time, a community well, pump on the valley floor, and water distribution system was installed Chiarella and Beck.

Israel

Researchers in Israel were the first to experiment in the application of new techniques to water harvesting. They have found various methods effective in increasing run off from land surfaces, either singly or in combination, including land smoothing and compaction, formation of sodic crusts spraying applications of various asphaltic materials. Among the asphaltic formulations, heavy fuel oil diluted with kerosene proved to be both effective and economical.

They found area ratios between contributing and receiving areas on the order of 3:1 to 6:1 in a rainfall zone of 200 to 250 mm. The accretion of water received in the planting zones of experimental plots provided soil moisture equivalent to the entire normal winter rainfall in the mediterranean climatic zone of the country, in northern Israel, where unirrigated orchards provide a livelihood for an appreciable number of farmers.

Gaps in Knowledge

The economics of water harvesting have never been fully documented, particularly over the long term, even in Israel, Australia and the United States where most of the modern technology of water harvesting has been developed. The need is for a better understanding of the economic viability of different methods in different economic environments, particularly in those of the developing countries.

A primary function of water harvesting system designed for livestock watering, domestic supplies, and run off farming, in addition to providing more useable water, is to smooth out the variation in natural rainfall by providing water during interrain periods. Never-the-less, the reliability of a system and the degree of risk associated with a system depends upon the reliability of rainfall and water demand. Thus, rainfall records are essential to design and operation. They are often insufficient in the developing countries and especially in the less dense populated arid zones of those countries. Furthermore, because of the high variation in rainfall from year to year in arid zones, records are needed from longer historical periods than would be required for similar analysis in temperate zones. More adequate coverage by weather stations and maintenance of daily records should be encouraged by international development agencies.

Information is badly needed on the minimum water requirements of agricultural crops in different climatic regions of the arid zones in order to reduce the risk in designing water harvesting systems for agriculture.

Evaporation from storage is a serious problem. Two meters or more water loss from free water surfaces is not uncommon in arid lands. As yet, except for tanks and cisterns, no economical and effective method has been developed to prevent evaporation from storage ponds despite the great variety of materials tried.

An important technical-research need is to reduce the costs of catchment treatment and to make the treatment practical for a wider variety of soils and situations. Industry is constantly formulating new materials which should be continually monitored and evaluated for use both for catchment treatment and in evaporation control.

The technology of water harvesting will largely be advanced through empirical experimentation. Thus, more information and hard data are needed from a wider range of climatic, soil, economic and social conditions. Information, of course, is needed on successful projects, but just as valuable and much more difficult to obtain is information on failures and the reasons thereof. Efforts should be made toward establishing an accessible international outlet for such information.

As more information becomes available, more work is needed on the modeling and synthesis of water harvesting systems. Until adequate prediction schemes are developed the design of water harvesting systems will remain dependent upon the experience of a limited number of experts who have learned the hard way.

Although there are a large number of techniques available and many variations within the techniques, new ideas are needed. for example, the use of plastic greenhouses, where water is continuously recycled, has been developed to the extent that it could be coupled with a water harvesting system to produce high value crops. This has not yet been done.

Quantitative information on the quality of water collected by water harvesting systems is limited. The information is needed for various catchment treatments to compare with safe standards for livestock use, human consumption and for growing plants. Water quality analysis should be an integral part of any water harvesting project.

Water harvesting offers a method of effectively developing the scarce water resources of arid regions. As contrasted to the development of groundwater, which is usually a finite water resource in arid zones, the method allows use of the renewable rainfall which occurs, even though in limited amounts, year in and year out. It is also a relatively inexpensive method of water supply that can be adapted to the resources and needs of the rural poor. It is necessarily small scale, and as such it can provide stability and improve the quality of life in small rural communities and that of small land holders who are several stages removed from the benefits of large scale development projects. Despite this, water harvesting is not a panacea. It involves some risk, dependent upon the vagaries of climate. New skills, though simple, are required, maintenance is a constant necessity, and good design is imperative.

References

Al-Labadi, M., "Water Harvesting in Jordan- Existing and Potential Systems", *In: FAO, Water Harvesting For Improved Agricultural Production*, Expert Consultation, Cairo, Egypt 21-25 Nov. 1993.

Bruins, H. J., Evenari, M. and Nessler, U., *Rainwater harvesting agriculture for food production in arid zones: The challenge of the African famine*, Appl. Geogr. 6:13-33, 1986.

Cullis Adrian, Pacey Arnold, *Rain Water Harvesting,* London, Intermediate Technology Publisher, 1991.

Ladda, Brijgopal, *Planning and Financing Urban Water Supply: A Case Study of Ahmedabad,* Ahmedabad, CEPT, 1995.

Siegert, K., "Introduction to Water Harvesting. Some Basic Principles for Planning", *Design and Monitoring. In: FAO, Water Harvesting For Improved Agricultural Production*, Expert Consultation, Cairo, Egypt 21-25 Nov. 1993.

4

Soil Water Conservation

Since 1950, agricultural policymakers have been confronted with a new and vexing set of problems. Water quality problems resulting from the presence of nutrients, pesticides, salts, and trace elements have been added to an historical concern for soil erosion and sedimentation. Economic problems in the 1980s intensified concern about the loss of family farms and rural development issues. Maintaining the ability of agriculture to compete in international markets became a central tenet of agricultural policy, and agriculture became a central issue in international trade talks. At the same time profound structural changes were occurring in the agricultural sector and new technologies were changing the face of agricultural production. The search for solutions to these different but related problems has dominated debate over agricultural policy.

Soil and water quality problems caused by agricultural production practices are receiving increased national attention and are now perceived by society as environmental problems comparable to other national environmental problems such as air quality and the release of toxic pollutants from industrial sources. Severe soil degradation from erosion, compaction, or salinization can destroy the productive capacity of the soil and exacerbate water pollution from sediment and agricultural chemicals. Sediments from eroded croplands interfere with the use of waterbodies for transportation; threaten investments made in dams, locks, reservoirs, and other developments; and degrade aquatic ecosystems. Nutrients accelerate the rate of eutrophication of lakes, streams, and estuaries; and nitrogen in the form of nitrates can cause health problems if ingested by humans in drinking water. Pesticides in drinking water can become a human health concern and have been suggested to disrupt aquatic ecosystems. Salts can be toxic at high enough levels and can seriously reduce the uses to which water can be put. In some areas, toxic trace elements in irrigation drainage water have caused serious damage to fish, wildlife, and aquatic ecosystems.

Importance of Soil Water Conservation

Both soil and water are essential for plant growth. The soil provides a structural base to the plants and allows the root system to spread and get a strong hold. The pores of the soil within the root zone hold moisture which clings to the soil particles by surface tension in the driest state or may fill up the pores partially or fully saturating with it useful nutrients dissolved in water, essential for the growth of the plants. The roots of most plants also require oxygen for respiration. Hence, full saturation of the soil pores leads to restricted root growth for these plants.

Since irrigation practice is essentially, an adequate and timely supply of water to the plant root zone for optimum crop yield, the study of the inter relation ship between soil pores, its water-holding capacity and plant water absorption rate is fundamentally important. Though a study in detail would mostly be of importance to an agricultural scientist, in this lesson we discuss the essentials which are important to a water resources engineer contemplating the development of a command area through scientifically designed irrigation system.

Soil is a heterogeneous mass consisting of a three phase system of solid, liquid and gas. Mineral matter, consisting of sand, silt and clay and organic matter form the largest fraction of soil and serves as a framework with numerous pores of various proportions. The void space within the solid particles is called the soil pore space. Decayed organic matter derived from the plant and animal remains are dispersed within the pore space. The soil air is totally expelled from soil when water is present in excess amount than can be stored. On the other extreme, when the total soil is dry as in a hot region without any supply of water either naturally by rain or artificially by irrigation, the water molecules surround the soil particles as a thin film.

In such a case, pressure lower than atmospheric thus results due to surface tension capillarity and it is not possible to drain out the water by gravity. The salts present in soil water further add to these forces by way of osmotic pressure. The roots of the plants in such a soil state need to exert at least an equal amount of force for extracting water from the soil mass for their growth. In the following sections, we discuss certain important terms and concepts related to the soil-water relations. First, we start with a discussion on soil properties and types of soils.

Properties of Soil

Soil is a complex mass of mineral and organic particles. The important properties that classify soil according to its relevance to making crop production (which in turn affects the decision making process of irrigation engineering) are:

— Soil texture

— Soil structure

Soil Texture

This refers to the relative sizes of soil particles in a given soil. According to their sizes, soil particles are grouped into gravel, sand, silt and day. The relative proportions of sand, silt and clay is a soil mass determines the soil texture.

According to textural gradations a soil may be broadly classified as:

— Open or light textural soils: these are mainly coarse or sandy with low content of silt and clay.

— Medium textured soils: these contain sand, silt and clay in sizeable proportions, like loamy soil.

— Tight or heavy textured soils: these contain high proportion of clay.

Soil Structure

This refers to the arrangement of soil particles and aggregates with respect to each other. Aggregates are groups of individual soil particles adhering together. Soil structure is recognized as one of the most important properties of soil mass, since it influences aeration, permeability, water holding capacity, etc. The classification of soil structure is done according to three indicators as:-

— *Type*: There are four types of primary structures-platy, prism-like, block like and spheroidal.

— *Class*: There are five recognized classes in each of the primary types. These are very fine, fine, medium, coarse and very coarse.

— *Grade*: This represents the degree of aggradation that is the proportion between aggregate and unaggregated material that results when the aggregates are displaced or gently crushed. Grades are termed as structure less, weak, moderate, strong and very strong depending on the stability of the aggregates when disturbed.

Classification of Soil

Soils vary widely in their characteristics and properties. In order to establish the interrelation ship between their characteristics, they need to be classified. In India, the soils may be grouped into the following types:

— *Alluvial soils*: These soils are formed by successive deposition of silt transported by rivers during floods, in the flood plains and along the coastal belts. This group is by for the largest and most important soil group of India contributing the greatest share to its agricultural wealth. Though a great deal of variation exists in the type of alluvial soil available throughout India, the main features of the soils are derived from the deposition laid by the numerous tributaries of the Indus, the Ganges and the Brahmaputra river systems. These streams, draining the Himalayas, bring with them the products of weathering rocks constituting the mountains, in various degrees of fineness and deposit them as they traverse the plains. Alluvial soils textures vary from clayey loam to sandy loam. The water holding capacity of these soils is fairly good and is good for irrigation.

— *Black soils*: This type of soil has evolved from the weathering of rocks such as basalts, traps, granites and gneisses. Black soils are derived from the Deccan trap and are found in Maharashtra, western parts of Madhya Pradesh, parts of Andhra Pradesh, parts of Gujarat and some parts of Tamilnadu. These soils are heavy textured with the clay content varying from 40 to 60 percent.the soils possess high water holding capacity but are poor in drainage.

— *Red soils*: These soils are formed by the weathering of igneous and metamorphic rock comprising gneisses and schist's. They comprise of vast areas of Tamil nadu, Karnataka, Goa, Daman & Diu, south-eastern Maharashtra, Eastern Andhra Pradesh, Orissa and Jharkhand. They also are in the Birbhum district of West Bengal and Mirzapur, Jhansi and Hamirpur districts of Uttar pradesh. The red soils have low water holding capacity and hence well drained.

— *Laterites and Lateritic soils*: Laterite is a formation peculiar to India and some other tropical countries, with an intermittently moist climate. Laterite soils are derived from the weathering of the laterite rocks and are well developed on the summits of the hills of the Karnataka, Kerala, Madhya Pradesh, The eastern ghats of Orissa, Maharashtra, West Bengal, Tamilnadu and Assam. These soils have low clay content and hence possess good drainage characteristics.

— *Desert soils*: A large part of the arid region, belonging to western Rajasthan, Haryana, Punjab, lying between the Indus river and the Aravalli range is affected by the desert conditions of the geologically recent origin. This part is covered by a mantle of blown sand which, combined with the arid climate, results in poor soil development. They are light textured sandy soils and react well to the application of irrigation water.

— *Problem soils*: The problem soils are those, which owing to land or soil characteristics cannot be used for the cultivation of crops without adopting proper reclamation measures. Highly eroded soils, ravine lands, soils on steeply sloping lands etc. constitute one set of problem soils. Acid, saline and alkaline soils constitute another set of problem soil.

Soil Water Classification

As stated earlier, water may occur in the soil pores in varying proportions. Some of the definitions related to the water held in the soil pores are as follows:

— *Gravitational water*: A soil sample saturated with water and left to drain the excess out by gravity holds on to a certain amount of water. The volume of water that could easily drain off is termed as the gravitational water. This water is not available for plants use as it drains off rapidly from the root zone.

— *Capillary water*: The water content retained in the soil after the gravitational water has drained off from the soil is known as the capillary water. This water is held in the soil by surface tension. Plant roots gradually absorb the capillary water and thus constitute the principle source of water for plant growth.

— *Hygroscopic water*: The water that an oven dry sample of soil absorbs when exposed to moist air is termed as hygroscopic water. It is held as a very thin film over the surface of the soil particles and is under tremendous negative (gauge) pressure. This water is not available to plants.

The above definitions of the soil water are based on physical factors. Some properties of soil water are not directly related to the above significance to plant growth.

Soil Water Constants

For a particular soil, certain soil water proportions are defined which dictate whether the water is available or not for plant growth. These are called the soil water constants, which are described below.

— *Saturation capacity*: This is the total water content of the soil when all the pores of the soil are filled with water. It is also termed as the maximum water holding capacity of the soil. At saturation capacity, the soil moisture tension is almost equal to zero.

— *Field capacity*: This is the water retained by an initially saturated soil against the force of gravity. Hence, as the gravitational water gets drained off from the soil, it is said to reach the field capacity. At field capacity, the macro-pores of the soil are drained off, but water is retained in the micropores. Though the soil moisture tension at field capacity varies from soil to soil, it is normally between 1/10 (for clayey soils) to 1/3 (for sandy soils) atmospheres.

— *Permanent wilting point*: Plant roots are able to extract water from a soil matrix, which is saturated up to field capacity. However, as the water extraction proceeds, the moisture content diminishes and the negative (gauge) pressure increases. At one point, the plant cannot extract any further water and thus wilts.

Two stages of wilting points are recognized and they are:

— *Temporary wilting point*: This denotes the soil water content at which the plant wilts at day time, but recovers during right or when water is added to the soil.

— *Ultimate wilting point*: At such a soil water content, the plant wilts and fails to regain life even after addition of water to soil.

It must be noted that the above water contents are expressed as percentage of water held in the soil pores, compared to a fully saturated soil.

Water Absorption by Plants

Water is absorbed mostly through the roots of plants, though an insignificant absorption is also done through the leaves. Plants normally have a higher concentration of roots close to the soil surface and the density decreases with depth as shown in Figure 1.

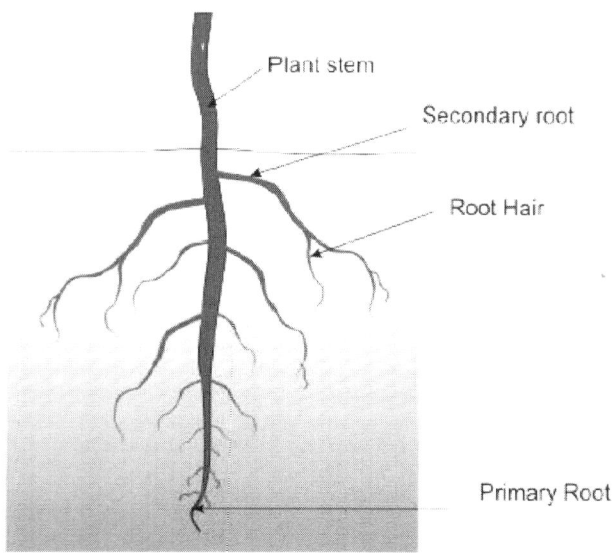

Figure 1. Typical root density variation of a plant with depth

In a normal soil with good aeration, a greater portion of the roots of most plants remain within 0.45m to 0.60m of surface soil layers and most of the water needs of plants are met from this zone. As the available water from this zone decreases, plants extract more water from lower depths. When the water content of the upper soil layers reach wilting point, all the water needs of plants are met from lower layers. Since there exists few roots in lower layers, the water extract from lower layers may not be adequate to prevent wilting, although sufficient water may be available there.

When the top layers of the root zone are kept moist by frequent application of water through irrigation, plants extract most of the water (about 40 percent) from the upper quarter of their root zone. In the lower quarter of root zone the water extracted by the plant meets about 30 percent of its water needs. Further below, the third quarter of the root zone extracts about 20 percent and the lowermost quarter of root zone extracts the remaining about 10 percent of the plants water.

It may be noted that the water extracted from the soil by the roots of a plant moves upwards and essentially is lost to the atmosphere as water vapours mainly through the leaves. This process, called transpiration, results in losing almost 95 percent of water sucked up. Only about 5 percent of water pumped up by the root system is used by the plant for metabolic purpose and increasing the plant body weight.

Watering Intervals

A plot of land growing a crop has to be applied with water from time to time for its healthy growth. The water may come naturally from rainfall or may supplemented by artificially applying water through irrigation. A crop should be irrigated before it receives a set back in its growth and development. Hence the interval between two irrigations depends primarily on the rate of soil moisture depletion. Normally, a crop has to be irrigated before soil moisture is depleted below a certain portion of its availability in the root zone depending on the type pf plant.

The intervals are shorter during summer than in winter. Similarly, the intervals are shorter for sandy soils than heavy soils. When the water supply is very limited, then the interval may be prolonged which means that the soil moisture is allowed to deplete below 50 percent of available moisture before the next irrigation is applied. The optimum rates of soil moisture for a few typical crops are given below.

— *Maize*: Field capacity to 60 percent of availability

— *Wheat*: Field capacity to 50 percent of availability

— *Sugarcane*: Field capacity to 50 percent of availability

— *Barley*: Field capacity to 40 percent of availability

— *Cotton*: Field capacity to 20 percent of availability

As for rice, the water requirement is slightly different than the rest. This is because it requires a constant standing depth of water of about 5cm throughout its growing period. This means that there is a constant percolation of water during this time and it has estimated that about 50 to 70 percent of water applied to the crop is lost in this way. For most of the crops, except rice, the amount of water applied after each interval should be such that the moisture content of the soil is raised to its field capacity.

The soil moisture depletes gradually due to the water lost through evaporation from the soil surface and due to the absorption of water from the plant roots, called transpiration more of which has been discussed in the next session. The combined effect of evaporation and transpiration, called evapo-transpiration (ET) decides the soil water depletion rate for a known value of ET (which depends on various factors, mainly climate); it is possible to find out the irrigation interval. Some of the operational soil moisture ranges of some common crops are given below:

Rice

This crop is grown both in lowland and upland conditions and throughout the year in some parts of the country. For lowland rice, the practice of keeping the soil saturated or upto shallow submergence of about 50mm throughout the growing period has been found to be the most beneficial practice for obtaining maximum yields. When water resources are limited, the land must be submerged atleast during critical stages of growth. The major portion of the water applied to the rice crop, about 50-75% is lost through deep percolation which varies with the texture of the soil. Since the soil is kept constantly submerged for rice growth, all the pores are completely filled with water through it is in a state of continuous downward movement. The total water required by the rice plant is about 1.0 to 1.5m for heavy soils and soils with high water table; 1.5 to 2.0m for medium soils and 2.0 to 2.5 for light soils with deep water table.

Wheat

The optimum soil moisture range for tall wheats is from the field capacity to 50% of availability. The dwarf wheats need more wetness, and the optimum moisture range is from 100 to 60 percent availability. The active root zone of the crops varies from 0.5 to 0.75m depending upon the soil type. The total water requirement for wheat plants vary from 0.25m to 0.4 m in northern India to about 0.5m to 0.6m in Central India.

Barley

This crop is similar to wheat in its growing habits, but can withstand more droughts because of the deeper and well spread root system. The active root zone of Barley extends between 0.6m to 0.75m on different soil types. The optimum soil moisture ranges from the field capacity to 40% of availability.

Maize

The crop is grown almost all over the country. The optimum soil moisture range is from 100 to 60% of availability in the maximum root zone depth which extends from 0.4 to 0.6 on different soil types. The actual irrigation requirement of the crop varies with the amount of rainfall. In north India, 0.1m and 0.15m is required to establish the crop before the onset of monsoon. In the south, it is found that normal rain fall is sufficient to grow the crop in the monsoon season where as 0.3m of water is required during water.

Cotton

The optimum range of soil moisture for cotton crop is from the field capacity to 20% of available water. He root zone varies upto about 0.75m. The total water requirement is about 0.4m to 0.5m.

Sugarcane

The optimum soil moisture for sugarcane is about 100 to 50 percent of water availability in the maximum root zone, which extends to about 0.5m to 0.75m in depth. The total water depth requirement for sugarcane varies from about 1.4m to 1.5m in Bihar; 2.2m- 2.4m in Karnataka; and 2.0 - 2.3m in Madhya Pradesh.

Significance of Water in Plants

During the life cycle of a plant water, among other essential elements like air and fertilizers, plays a vital role, some of the important ones being:

— Water maintains the turgidity of the plant cells, thus keeping the plant erect. Water accounts for the largest part of the body weight of an actively growing plant and it constitutes 85 to 90 percent of the body weight of young plants and 20 to 50 percent of older or mature plants.

— Water provides both oxygen and hydrogen required for carbohydrate synthesis during the photosynthesis process.

— Water acts as a solvent of plant nutrients and helps in the uptake of nutrients from soil.

— Food manufactured in the green parts of a plant gets distributed throughout the plant body as a solution in water.

— Transpiration is a vital process in plants and does so at a maximum rate (called the potential evapo transpiration rate) when water is available in adequate amount. If soil moisture is not sufficient, then the transpiration rate is curtailed, seriously affecting plant growth and yield.

— Leaves get heated up with solar radiation and plants help to dissipate the heat by transpiration, which itself uses plant water.

Irrigation Agriculture

In irrigation agriculture, the quality of water used for irrigation should receive adequate attention. Irrigation water, regardless of its source, always contains some soluble salts in it. Apart from the total concentration of the dissolved salts, the concentration of some of the individual salts, and especially those which are most harmful to crops, is important in determining the suitability of water for irrigation.

The constituents usually determined by analysing irrigation water are the electrical conductivity for the total dissolved salts, soluble sodium percentage, sodium absorption ratio, boron content, pH, cations such as calcium, magnesium, sodium, potassium and anions such as carbonates, bicarbonates, sulphates, chlorides and nitrates.

Water from rivers which flow over salt effected areas or in the deltaic regions has a greater concentration of salts sometimes as high as 7500 ppm or even more. The quality of tank or lake water depends mainly on the soil salinity in the water shed areas and the aridity of the region. The quality of ground water resources, that is, from shallow or deep wells, is generally poor under the situations of:

— high aridity

— high water table and water logged conditions

— in the vicinity of sea water

Some Definitions

1. *Root Zone*: The soil root zone is the area of the soil around the plant that comes in contact with the plant root (Figure 2).

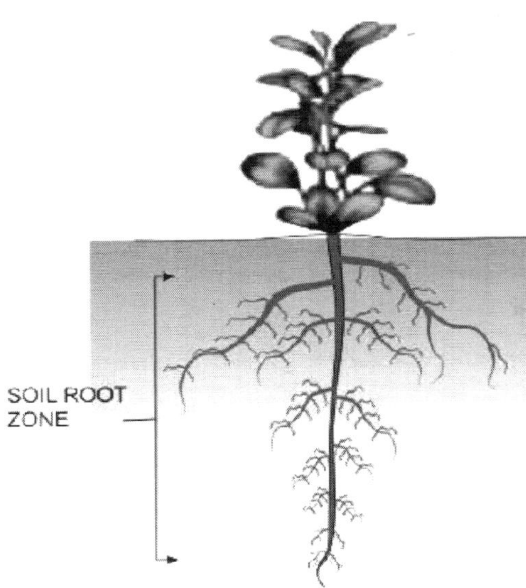

Figure 2. Definition of soil root zone

2. *Soil Moisture tension*: In soils partially saturated with water there is moisture tension, which is equal in magnitude but opposite in sign to the soil water pressure. Moisture tension is equal to the pressure that must be applied to the soil water to bring it to a hydraulic equilibrium, through a porous permeable wall or membrane, with a pool of water of the same composition.
3. *Wilts*: Wilting is drooping of plants. Plants bend or hang downwards through tiredness or weakness due to lack of water.

Hazard of Soil Erosion

Soil erosion is the wearing away of land surface by the action of such natural agencies as water and wind. In India, there is very little area free from the hazard of soil erosion. It is estimated that out of 305.9 million hectares of reported area, 145 million hectares is in need of conservation measures. Severe erosion occurs in the sub-humid and per-humid areas due to high rainfall and improper management of land and water. Agricultural land in the major part of the country suffers from erosion. Apart from reducing the yields through the loss of nutrients, erosion destroys the soil resources itself every year. For example, in Maharashtra over 70 per cent of the cultivated land has been affected by erosion in varying degrees and 32 per cent of the land having been highly eroded is no longer cultivable.

In the Sholapur district, nearly 17 per cent of the land of medium depth (more than 45 cm) has deteriorated into shallow soils (less than 45 cm) in 75 years from 1870 to 1945. Similarly, in Akola, Buldana and Yeotmal districts, the number of fields with less than 37.5 cm soil depth increased during the same period by 54, 16 and 8 per cent respectively. As much as 2.3 million ha is already under ravines scattered all over India. The ravines apart from ruining the soil resources for ever are a constant threat to the adjoining fertile cultivated lands.

The denudation of forests and vegetation in the Shivalik Hills, the Himalayas, the Western Ghats, the Eastern Ghats and other mountain ranges of the Deccan have resulted in floods and chos (rainy season torrents) which destroy good agricultural land. For instance, the chos in the Hoshiarpur district of Punjab covered an area of 19,282 ha which increased to 32,022 ha in 1884, to 37,730 ha in 1897 and to about 60,000 ha in 1936. In the Himalayan regions, landslides and landslips are very serious problems caused by improper land management.

The recent landslide disaster in the winter of 1969 in the Darjeeling district of West Bengal is a reminder of the bad management of land resources and a portent of worse things to come. The erosion problem along the mountain roads is assuming very serious proportions. The Border Road Organization is finding this problem so acute that a National Seminar was organized on the problems of controlling erosion and stabilizing slopes along the highways. A huge amount of money is spent every year for keeping these important lines of communication open.

Costly reservoirs constructed under the river-valley projects are being silted up at an alarming rate owing to the denudation of forest vegetation, the cultivation of steep slopes without adopting any conservation practices, landslides, torrents, etc. Morever, as the pressure on land increases there will be a tendency and a demand for opening up marginal and steep lands for cultivation. These lands will be in greater need for measure to conserve soil and water. Owing to our present status as a developing country we are not yet faced with the problem of environmental pollution, though the pollution of soil, water and atmopsphere is round the corner as we increase the pace of our development and exploit theses resources.

Sediment load is certainly one of our greatest agricultural hazards, particularly in the case of rivers and canals. In a country, such as USA, even after 35 years of efforts, and expenditure of large sums there has been little, if any, reduction in the overall sediment load in the US streams, mainly owing to the increase in her non-agricultural activities. In India, the situation is infinitely worse, since the sediment load from agricultural lands not only continue unabated, but it is also on the increase and the sources of sediment are multiplying because of the fast rate of our developmental activities. In fact, if erosion is permitted to continue at its present rate, it is possible that all work will be the reclamation of soil rather than the conservation and management of soil and water.

Types of Soil Erosion

The following types of erosion are observed in general:

1. *Normal or geologic erosion*: This is a normal feature of any landscape. Geologic erosion takes place steadily but so slowly that ages are required for it to make any marked alteration in the major features of the earth's surface. There is always an equilibrium between the removal and formation of soil, so that unless the equilibrium is disturbed by some outside agency, the mature soil preserves more or less, a constant depth and character indefinitely.
2. *Accelerated soil erosion*: The removal of the surface soil from areas denuded of their natural protective cover as a result of human and animal interference takes place at a much faster rate than that at which it is built up by the soil-forming process. This accelerated detachment rapidly ravages the land and it is with this type of soil erosion that we are so seriously concerned. Nature requires, on an average, about 1000 years to build up 2.5 cm of top soil, but wrong farming methods may take only a few years to erode it from lands of average slope.
3. *Wind erosion*: Wind erosion takes place normally in arid and semi-arid areas devoid of vegetation, where the wind velocity is high. The soil particles on the land surface are lifted and blown off as dust storms. When the velocity of the dust-bearing winds is retarded, coarser soil particles are deposited in the form of dunes and thus fertile lands are rendered unfit for cultivation. In other places, fertile soil is blown away by winds and the subsoil is exposed, as a result the productive capacity of the soil is considerably reduced.

4. *Water erosion*: Soil erosion caused by water can be distinguished in three forms, viz. (1) sheet erosion, (2) rill erosion, and (3) gully erosion.
5. *Sheet erosion*: Sheet erosion removes a thin covering of soil from large areas, often from entire fields, more or less, uniformly during every rain which produces a run-off. This type of erosion is very insidious, since it keeps the cultivator almost ignorant of its ill-effects. It is generally neglected, although the soil deteriorates slowly and imperceptibly. Its existence, however, can be detected by the muddy colour of the run-off from the fields.
6. *Gully erosion*: When rill erosion is neglected, the tiny grooves develop into wider and deeper channels, which may assume a huge size. This is called 'gully' erosion. Gullies are the most spectacular evidence of the destruction of soil. The gullies tend to deepen and widen with every heavy rainfall. They cut up large fields into small fragments and, in course of time, make them unfit for cultivation.
7. *Landslides or slip erosion*: A landslide is defined as an outward and downward movement of the slope-forming material, composed of natural rocks, soil, artificial fills, etc. The fundamental causes of landslides are topography of the region and geological structure, the kinds of rocks and their physical characteristics. The immediate cause of a slide may be an earthquake, or a heavy rainfall, which unduly saturates the ground or a part of a road. However, these are accidents rather than fundamental causes.
8. *Stream-bank erosion*: Torrents are defined as hill streams characterized by wide-spreading beds on emergence from the hills with ill-defined banks, flashy flows and swift currents. Usually, they are dry water-courses, except during the rainy season when with every downpour in their catchment, they get very much swollen with flood and subside almost to its normal tiny size immediately after the storm is over.

These sudden and violent flows are responsible for moving immense quantities of detritus, comprising boulders, shingles, sand and silt, depending upon the geology of the terrain. This debris gets deposited in the torrent bed in the form of scattered islands owing to the sudden widening of the torrent channel after it emerges from the hills, or owing to the flattening of the gradient in the lower reaches, or because of obstructions caused by wild vegetation and uprooted trees. The bed level of the torrent is raised by these deposits. These deposits, in turn, reduce the transporting capacity of the torrent, resulting in overflows and the meandering of the course and in the erosion of the banks.

Mechanism of Erosion

Water Erosion

Soil erosion caused by rainfall is the result of the application of energy from two distinct sources, namely (i) the falling rain drops, and (ii) the surface flow. The energy of a falling rain drop is applied slantingly or vertically from above, whereas that of a surface flow is applied more or less horizontally along the surface of the ground. The chief role of the falling rain drop is to detach soil particles, whereas that of the surface flow is to transport the soil.

The falling rain drop also makes a major contribution to the movement of the soil on unprotected sloping lands during the period of heavy-impact storms, by splashing large quantities downslope and by imparting transporting capacity to the surface water by keeping it turbid. More than 100 tonnes of soil per hectare can sometimes be lost yearly in this fashion from a bare and highly detachable soil on slopping land.

Wind Erosion

Wind is responsible for three types of soil movement in the process of wind erosion. They are known as:

(i) saltation,

(ii) suspension, and

(iii) surface creep.

Saltation

The major portion of the soil carried by the wind is moved in a series of short bounces called "saltations". The soil carried in a saltation consists of fine particles ranging from 0.1 to 0.5 mm in diameter. saltation is caused by the direct pressure of wind on soil particles and their collision with other particles. After being pushed along the ground surface by the wind, the particles leap almost vertically in the first stages of saltation. Some grains rise only a short distance; others leap 30 cm or higher, depending on the velocity of the rise from the ground.

Suspension

Very fine soil particles, less than 0.1 mm in diameter, are carried into suspension, being kicked up into the air by the action of particles in saltation. The movement of fine dust in suspension is completely governed by the characteristic movement of the wind. Suspended material is carried long distances from its original location and is thus a complete loss to the eroded area, especially when erosive winds are from different directions.

Surface Creep

Soil particles, larger than about 0.5 mm in diameter but smaller than 0.1 mm, are too heavy to be moved in saltation but are pushed or spread along the surface by the impact of particles in saltation to form a surface creep. About 90 per cent of the total soil movement in wind erosion is below the height of 30 cm. and about 50 per cent of it is within 5 cm of the ground level. The control of wind erosion is mainly based on the reduction or elimination of movement in saltation.

Factors Influencing Soil Erosion

Soil erosion by water is influenced greatly by:

(i) precipitation (its intensity and amount),

(ii) the slope of the land (its degree and length),

(iii) the type of soil, and

(iv) the nature of the ground cover and land use.

Precipitation

Precipitation is the most important factor influencing soil erosion. The intensity of rainfall, its duration and frequency influence the rate and the volume of run-off. A light rain which can be easily absorbed in the soil causes no run-off and soil loss. As the intensity of rain increases and more rain falls than can enter the soil (infiltration), there is of run-off and soil loss.

Rainfall of long duration and greater frequency increases both the total run-off and soil loss. Apart from the intensity and duration of rainfall, the soil moisture is also important in determining the run-off and soil loss by erosion. If the soil is already saturated with water, the same amount and intensity of rainfall will cause more run-off and soil loss from it than from a dry soil.

Slope of the Land

The speed and the extent of run-off depend on the slope of the land. The greater the slope, the greater is the velocity of the flow of the run-off. According to the law of falling bodies, velocity varies as the square-root of the vertical drop. Hence, if the land slope is increased four times, the velocity of the water flowing on the slope is approximately doubled. If the velocity of the run-off water is doubled, its energy, i.e. erosive power, is increased four times, as the latter varies as the square of the velocity.

Similarly, the quantity of the material of a given size that can be carried is increased about thirty-two times (varies as the fifth power of the velocity), and the size of the particles that can be transported by pushing or rolling is increased about sixty-four times (varies as the sixth power of the velocity). There is thus a rapidly increasing rate of soil loss as the slope of the field becomes steeper. The erosion hazard is not simply added but is multiplied as the field extends back on the steeper part of the farm.

Type of Soil

The type of soil, i.e. structure, texture, organic matter content, its infiltration capacity and permeability, greatly affects the soil loss and run-off. Fine soils are more susceptible to erosion than coarse soils, since rain-water enters in and passes through a dense clay much more slowly than through a porous sand or gravelly soil. In India, it has been observed that deep lateritic soils at Ootacamund and red soils at Deochanda have the lowest rate of run-off; the alluvial soils at Vasad and Dehra Dun have a very high rate of run-off; the black soils have an intermediate rate of run-off, but still the rate of run-off is high. Lateritic clays are less erodiable. The soil left in loose and pulverized condition is particularly liable to erosion through sheet-wash and gullying.

Nature of the Ground Cover

When rain falls on a surface covered by a thick mantle of plants, its velocity and erosive power are reduced and most of the water either quickly percolates through the soil or moves over the surface with non-erosive velocity. Areas not protected with thick cover of plants are unable to absorb water effectively, because the dashing rains shatter the soil surface, the fine soil particles go into suspension and the thick mixture of water and soil quickly fills and closes the tiny interstices in the soil, reducing infiltration and consequently increasing run-off and soil loss. Alongwith the loss of run-off of water and soil, considerable amount of plant nutrients are also lost. The loss of plant nutrients in alluvial soil at Kanpur increased with the increase in the degree of the slope and the increase was very steep when the degree of slope increased from 1.5 per cent to 3.0 per cent.

Land capability Classification

The key to soil and water conservation is the utilization and treatment of land according to its capability.

Any soil and water-conservation project includes two distinct sets of operations, viz. (1) the mapping of land for classification according to its capability, and (2) planning and executing measures to check erosion, improve land productivity and reclaim wasteland. The farm plans for effective soil and water conservation are based largely on the capability of the land. The land-capability classification map is normally prepared by interpreting a standard soil-survey map.

Land-capability classification is a systematic arrangement of different kinds of lands according to those properties that determine the ability of the land to produce crops on a virtually permanent basis. The factors determining land-capability. These are the major soil characteristics of the land, e.g. the texture of the top soil, its effective depth, permeability of the top soil and subsoil, and associated land features, e.g. the slope of the land, the extent of erosion, the degree of wetness and susceptibility to overflowing and flooding. The grouping of soils into capability classes is done primarily on the basis of their capability to produce common cultivated crops and pasture plants without deterioration over a long period.

The land-capability classes are based on the intensity of hazards and the limitations of use. The land-capability classes range from the best and most easily farmed land to that which has no value for cultivation, grazing or forestry, but which may be suited to wild-life, recreation or for watershed protection. They all fall into 2 broad groups : one suitable for cultivation and other land uses, and the other not suitable for cultivation, but suitable for other land uses.

Class I (Green Colour)

Soils in class I have very few or no limitations that restrict their use. This type of land is nearly level and the erosion hazard is low. The soils are deep, well-drained, easily worked, hold water well and are either fairly well supplied with plant nutrients or highly responsive to the application of fertilizers. The soils are not subject to damage because of overflow. The local climate must be favourable for growing many of the common field crops.

In irrigated areas, the soils may be in class I, if the limitation of the arid climate has been removed by relatively permanent irrigation works. These soils need ordinary management practices to maintain productivity. Such practices may include the use of one or more of the following: fertilizers, lime, cover and green-manure crops, conservation of crop residues and crop rotations. Soils in this class are suited to a wide range of plants, may be used for cultivated crops, pastures, forests, and wildlife, food and cover.

Class II (Yellow Colour)

Soils in class II have some limitations which reduce the choice of plants or require simple conservation practices. The limitations of soils in class II may result from the effects of one or more of the following factors:

(i) a gentle slope,

(ii) a slight susceptibility to erosion,

(iii) less than ideal soil depth,

(iv) occasional damaging overflow,

(v) wetness which can be corrected by drainage, but existing permanently as a moderate limitation,

(vi) slight to moderate salinity or sodium, easily corrected but likely to recur, and

(vii) a slight climatic limitation on soil use and management.

These soils require careful management. The limitations are only a few and the practices are easy to apply. They may need one or more of the following practices: terracing, strip cropping, contour cultivation, water disposal area, covered with vegetation crop rotation, cover and green-manure crops, stubble mulching, the use of fertilizers, manure and lime.

These soils may be used for growing cultivated crops, raising pastures, forests, and for wild-life, food and cover.

Class III (Red Colour)

Soils in class III have moderate limitations which reduce the choice of plants or require special conservation practices. Soils in class III have more restrictions than those in class II and, when used for cultivated crops, the conservation practices are usually more difficult to apply and to maintain.

Limitations of soils in class III may result from the effects of one or more of the following factors:

(i) a moderately sloping land,

(ii) moderately susceptibility to water or wind erosion,

(iii) frequent overflow accompanied with some crop damage,

(iv) very slow permeability of the sub-soil,

(v) wetness or continuing water-logging after drainage,

(vi) shallow soil depth up to the bed-rock, hard-pan or clay-pan which limits the rooting-zone and water storage,

(vii) low moisture-holding capacity,

(viii) moderate salinity or sodium, and

(ix) moderate climatic limitations.

Class IV (Blue Colour)

Soils in class IV have severe limitations that restrict the choice of plants and require careful management. The restrictions in the use of these soils are greater than those in class III and the choice of plants is more limited. When these soils are cultivated, very careful management is required and the conservation practices are more difficult to apply and to maintain.

The use of these soils for cultivated crops is limited as a result of the effect of one or more of the permanent features, such as (i) steep slopes, (ii) severe susceptibility to water or wind erosion, (iii) severe effect of past erosion, (iv) shallow soil, (v) low moisture-holding capacity, (vi) frequent overflow accompanied with severe crop damage, (vii) excessive wetness or continuing hazard of water-logging after drainage, (viii) severe salinity or sodium, and (ix) moderately adverse climate. These soils can be used for crops, pastures, forests, and wild-life food and cover.

Class V (Dark Green or Uncloured)

Soils in class V have little or no erosion hazard, but have other limitations, the removal of which is not practicable. They are used largely for pastures, forests, and wild-life food and cover. Such land is nearly level and is not subject to more than slight wind or water erosion. Cultivation is not feasible because of one or more limitations, such as overflow, stoniness, wetness or severe climate.

Examples of class V land are (i) soils of lowlands subject to frequent overflows which prevent the normal production of cultivated crops, (ii) nearly level soils with growing season that prevents the normal production of cultivated crops, (iii) the level or nearly level stony or rocky soils, and (iv) ponded areas where drainage for cultivated crops is not feasible but where soils are suitable for grasses or trees. Soils in class V are not suitable for raising cultivated crops, but are suitable for perennial vegetation (grazing and forestry, with few or no limitation). Pastures can be improved, and benefits from proper management can be expected. Physical conditions of soils are such that it is practicable to apply pasture improvements, if needed, such as seeding, liming, fertilizing, and water control with contour furrows, drainage ditches, diversions of water spreaders.

Class VI (Orange Colour)

Soils in class VI have severe limitations that make them unsuitable for cultivation and limit their use largely for pastures, or forests, or wild-life food and cover.

Soils in class VI have continuing limitations which cannot be corrected, such as (i) steep slope, (ii) very severe erosion hazard, (iii) very severe effect of past erosion (iv) stoniness, (v) shallow rooting-zone, (vi) excessive wetness or overflow, (vii) low moisture capacity, (viii) salinity or sodium, and (ix) severe climate. Soils in this class are subject to moderate limitations under grazing or forestry use.

Class VII (Brown Colour)

Soils in class VII have very severe limitations that make them unsuitable for cultivation and restrict their use largely to grazing, or forestation, or wild-life food and cover. The soils in this class are subject to severe limitations or hazards under either grazing or forestry use. The physical condition of soils is such that it is not practicable to adopt pasture improvements and water-control practices. Soil restrictions are severe than those in the case of class VI soils.

Class VIII (Purple Colour)

Soils and land forms in class VIII have limitations that preclude their use for commercial plant production and restrict their use to recreation, wild-life food and cover or to water-supply, water shed protection or for aesthetic purposes. Significant return on site benefits from soils and land forms in class VIII cannot be expected from management of crops, grasses or trees, although indirect benefits from wild-life, watershed protection or recreation may be possible.

Limitations which cannot be corrected may result from the effects of one or more of the following factors: (i) erosion or erosion hazard, (ii) severe climate, (iii) wet soil, (iv) stones, (v) low moisture capacity, (vi) salinity or sodium. Bad lands, rock outcrops, sandy beaches, marshes, deserts, river wash, mine tailings and other nearly barren lands are included in class VIII in order to protect other more valuable soils to control water or for wild-life or for aesthetic reasons. The land-capability class is indicated on the maps by roman numerals I to VIII or by standard colours or by both.

Land-capability Subclass Level

A land-capability class is determined by the degree of limitations in land use together with the hazards involved. For example, in class III land, we have land suitable for cultivation but subject to moderate hazard of water erosion because of a steep slope or a moderate hazard of wind erosion on smooth land or moderate hazard of water-logging or overflow and a shallow depth to the bed-rock. Each of these kinds of limitations are recognized at the subclass. Four kinds of limitations are recognized at the subclass level.

Subclass E (Erosion)

It is made up of soils where the susceptibility to erosion is the dominant hazard in their use. Susceptibility to erosion and damage due to past erosion are major factors that govern the placing of soils in this class.

Subclass W (Excess water)

It is made up of soils where excess water is the dominant hazard in their use. Poor soil drainage, wetness, high water-table and overflow are the criteria for determining which soils belong in this subclass.

Subclass S (Soil Limitations within the rooting zone)

It includes soils which have such limitations as the shallowness of the rooting-zone, stones, low moisture-holding capacity, low fertility difficult to correct and salinity or alkalinity.

Subclass C (Climatic limitations)

It includes soils where the climate (temperature or lack of moisture) is the only major hazard of limitations in their use. The land-capability subclass is designated by the small letters which follow the land-capability classes. Roman numerals, e.g. IIe, IIIs, IVw capability class I has no subclass. Where two kinds of limitation can be modified or corrected and are essentially equal, the subclasses have the following priority e, w, s, c. where soils have two kinds of limitations, the dominant one is shown first.

Practical Methods of Soil and Water Conservation

Broadly speaking, the practical methods of soil and water conservation fall into two important classes, viz. Agronomic measures and mechanical measures.

Agronomic Practices

Agronomic practices for soil and water conservation help to intercept rain drops and reduce the splash effect, help to obtain a better intake of water rate by the soil by improving the content of organic matter and soil structure, help to retard and reduce the overland run-off through the use of contour cultivation, mulches, dense-growing crops, strip-cropping and mixed cropping.

Contour-farming

During intense rain storms, the soil cannot absorb all the rain as it falls. The excess water flows down the slope under the influence of gravity. If farming is done up and down the slope, the flow of water is accelerated, because each furrow serves as a rill. The major part of the rain is drained away without infiltrating into the soil. The top fertile soil, along with plant nutrients and seeds, is washed off. All this results in a scanty and uneven growth of a crop.

A simple practice of farming across the slope, keeping the same level, as far as possible (which is technically called contour-farming) has many beneficial effects. The ridges and the rows of the plants placed across the slope form a continual series of miniature barriers to the water moving over the soil surface. The barriers are small individually, but as they are large in number, their total effect is great in reducing run-off, soil erosion and loss of plant nutrients. It has been experimentally proved that contour-farming reduces run-off and prevents soil erosion as compared with the up-and-down cultivation in the major groups of soils in India, viz. Alluvial soils, black soils and deep lateritic soils.

Apart from conserving the water and soil, contour-farming conserves soil fertility and increases crop yields. Contour-farming on alluvial soil, 2.2 per cent slope at Kanpur has conserved 11.3 kg of N, 11.7 kg of P_2O_5, 44.4 kg of K_2O, 398.1 kg of CaO, 118.1.kg of MgO in one season alone. These nutrients converted into fertilizers amount to 56.5 kg of sulphate of ammonia, 70 kg of single super-phosphate and 74 kg of muriate of potash per hectare in one season.

Contour-farming has also given 490 kg of johar grain and 273 kg of johar stalks more than the up-and-down cultivation at Kanpur. Thus every mm of rain-water conserved by contour-farming gave 22.5 kg of johar grain and 12.5 kg of johar straw more than the up-and-down cultivation. It has been proved that much less power is required to be exerted by man, animals and machines, if the cultivation is done in the contour instead of up-and-down the slope. There is less wear and tear of the implements and the same job is done in less time when contour-farming is practiced.

On long slopes, bunding is usually done to reduce the length of the slope. These bunds will serve as a good guide for contour-farming. All the cultural operations have to be done parallel to these bunds. On gentle slopes (between 0.5 to 2.0 per cent) bunding may not be essential. Contour (the line passing through the points having the same level) guidelines can be marked with the help of a hand level. On uniform slopes, these lines are to be marked about 50 m apart. Farming is done parallel to these lines.

The establishing of contour-farming on undulating land (having many depressions and ridges) is somewhat tedious. The water from each furrow collects in the depressions and results in breaches. The depressions are required to be filled up by levelling or may be left under grass. Mulching Surface mulches are used to prevent soil from blowing and being washed away, to reduce evaporation, to increase infiltration, to keep down weeds, to improve soil structure and eventually to increase crop yields.

Inter-culture kills weeds and produces a five or seven cm thick soil mulch which helps to reduce evaporation from the top soil. It also breaks the surface crust which forms after each downpour. Studies on mulching carried out in India under rain-fed agriculture have concentrated on the measurement of crop responses rather than on the manner in which crop responses are influenced.

At Dehra Dun, mulching with maize residues did not significantly influence the yield of the succeeding yield crop.

At Kota, mulching, in general, gave a higher yield of wheat grain and stalk than no mulching.

At Bellary, mulching with different materials did not give encouraging results in increasing the yields of cotton and johar crops. Mulching with straw og Encap Esso mulch increased the infiltration rate, improved soil moisture and increased the yields of wheat, barley, gram and linseed, succeeding the maize crop on eroded soils at Ranchi.

Growing of crops which provide the maximum cover, reduce run-off and soil loss. Cultivated legumes, in general, furnish a better cover and hence better protection to cultivated land against erosion than ordinary cultivated crops. The crops and the cropping systems will naturally vary from region to region, depending on the soil and climatic conditions.

At Dehra Dun, cowpeas provided the maximum canopy with or without the application of phosphate fertilizer. This was followed by mung, urad and dhaincha.

At Vasad, cowpeas provided the best vegetative cover for the soils. Sunnhemp was the next best; mung and groundnut were effective to some extent.

At Kota, velvet bean formed the canopy effective to the maximum extent. Cowpea was the next best.

At Kanpur, mung served as the most effective canopy, followed by urad and guar. The application of phosphorus increased the canopy.

At Rehmankhera, cowpea formed the best canopy, followed by Styzolobium mung, groundnut, urad, moth, and soybean. Sunnhemp, mung and Styzolobium helped to conserve soil moisture well, leading to a higher yield in the case of the succeeding barley crop. All the legumes were equally effective in reducing the loss of soil and nutrients.

Thus among the legumes, cowpea and mung proved to be important crops for providing a good cover for the land during the rainy season.

Bidi tobacco, being a clean-cultivated crop, resulted in the maximum soil loss and water loss. The soil and water losses were very much reduced, especially the water loss, when a cover-cum-manure crop was grown during the early monsoon before tobacco was transplanted. The inclusion of a cover-cum-green-manure crop increased the yield of tobacco from 1,585 to 1,850 kg/ha and reduced the need for nitrogenous fertilizer dose by 45%. Bajra+mung strip-cropping also reduced the water loss considerably.

Groundnut grown under two soil-climate conditions has been shown to reduce both the soil and water loss. Urad grown under two soil-climate conditions has been shown to be a run-off and soil-loss-permitting crop like maize. Intercropping maize with urad or arhar did not reduce the loss of soil and water.

Strip-cropping

Strip-cropping is essentially another form of rotation. Its importance in controlling the run-off erosion and thereby maintaining the fertility of the soil is now universally recognized. Strip-cropping, in effect, employs several good farming practices, including crop rotation, contour cultivation, proper tillage stubble-mulching, cover-cropping, etc. Strip-cropping is of the following different forms:

(i) Contour strip-cropping
(ii) Field strip-cropping
(iii) Wind strip-cropping
(iv) Permanent or temporary buffer strip-cropping.

Contour strip-cropping

Contour strip-cropping is the growing of soil-exposing and erosion-permitting crop in strips of suitable widths across the slopes on contour, alternating with strip of soil-protecting and erosion-resisting crops. Contour strip-cropping shortens the length of the slope, checks the movement of run-off water, helps to desilt it and increases the absorption of rain-water by the soil.

Further, the dense foliage of the erosion-resitant crops prevents the rain from beating the soil surface directly. It is advisable to rotate the strip-planting by sowing a non-resistant crop, following an erosion-resistant crop and vice versa. Very little experimental work has been done to determine the best combination of crops and other factors in a strip-cropping programme. The data collected at the Dry Farming Research Station, Sholapur, Maharashtra, have given the following indications:

(i) Groundnut, Moth bean (Phaseolus acontifolius) and Horse gram (Dolichos biflorus) are the most efficient and suitable crops for checking erosion.

(ii) The normal seed-rates of leguminous crops, other than groundnut, do not give sufficiently dense canopies to prevent rain drops from beating the soil surface in much cases. The seed-rate should be trebled.

(iii) The most effective width of the contour strips for cereals, such as jowar and bajra, is 21.6 m and for the intervening legume 7.2 m

Field strip-cropping

It is the planting of farm crops in more or less parallel strips across fairly uniform slopes, but not on exact contours.

Wind strip-cropping

It consists of planting tall-growing crops such as jowar, bajra or, maize, and low-growing crops in alternately arranged straight and long, but relatively narrow, parallel strips laid out right across the direction of the prevailing wind, regardless of the contour.

Permanent or temporary buffer strip-cropping

In the case of permanent or temporary buffer strip-cropping, the strips are established to take care of critical, i.e. steep or highly eroded, slopes in fields under contour strip-cropping. These strips do not form part of the rotation practiced in normal strip-cropping, and they are generally planted with perennial legumes, grasses or shrubs on a permanent or temporary basis.

Mixed cropping

The main reason for the lack of interest in strip-cropping in India is the small size of the holdings of the farmers. Instead, the Indian farmers have been interested in mixed farming which is very extensively adopted by him. Some of the important objectives of mixed cropping are a better and continuous cover of the land, good protection against the beating action of the rain, almost a complete protection against soil erosion and the assurance of one or more crops to the farmer. The roots of various species in a mixed crop feed at different depths in the soil.

Improvements in the practice of mixed cropping tend towards the rationalization of the mixtures with respect to composition and content and towards better arrangements than from the usual medley of crops. The line-sowing of mixed crops gives rise to the practice of intercropping. Out of five years of trial, the wheat crop failed in one year owing to the extreme drought, and gram failed in one year because of the incidence of gram wilt.

In both these years, the other mixed crops was a success, thus ensuring at least one crop to the farmer when mixed cropping was practiced. The wheat+gram mixture gave a consistently higher yield than the pure crops individually. The increase in net profit as a result of mixed cropping was 89 per cent and 183 per cent over the pure crops of wheat and gram, respectively.

Mechanical Measures

Mechanical measures play a very vital role in controlling erosion on agricultural land. They are adopted to supplement the agronomical practices when the latter alone are not adequately effective. These measures include basin-listing, subsoiling, contour-bunding, graded bunding and bench-terracing on steep slopes.

The main objectives of the mechanical measures for controlling erosion are: (i) to increase the time of concentration by intercepting the run-off and thereby providing an oppurtunity for the infiltration of water, and (ii) to divide a long slope into several short ones so as to reduce the velocity of the run-off and thus prevent erosion.

Basin-listing

Basin-listing consists in making of small interrupted basins along the counter with a special implement, called a basin-lister. Basin-listing helps to retain rain-water as it falls and is specially effective on retentive soils having mild slopes.

Subsoiling

This method consists in breaking with a subsoiler the hard and impermeable subsoil to conserve more rain-water by improving the physical conditions of a soil. This operation, which does not involve soil inversion and promotes greater moisture penetration into the soil, reduces both run-off and soil erosion. The subsoiler is worked through the soil at a depth of 30-60 cm at a spacing of 90-180 cm.

Countour Bunding

This practice consists in making a comparatively narrow-based embankment at intervals across the slope of the land on a level that is along the contour. It is an important measure that conserves soil and water in arid and semi-arid areas with high infiltration and permeability, and is commonly adopted on agricultural land up to a slope of about 6 per cent. As no specifications for bunding in deep black soil are available so far, no large-scale bunding on them can be recommended, pending further research.

In the alluvial soils of Gujarat, a vertical interval of 1.83 m and a cross-section of 1.3 sq m were found to be suitable for land with slopes ranging from 6-12% and for slopes less than 6% contour bunds with cross-sections of 0.9 to 1.3 sq m spaced at 0.9 to 1.2 m vertical interval were found to be effective. Observations and experiments at Sholapur, Bellary and Kota have shown that even in semi-arid climate, contour-bunding in deep black soils, with montmorillonitic type of clay is not suitable. The low rates of permeability and infiltration in these soils cause a prolonged impounding of water on the upstream side, and the crops are consequently damaged.

Graded Bunding

Graded bunding has been used in areas receiving rainfall of more than 80 cm per year, irrespective of the soil texture. In clay soils, graded bunding has to be used even for areas having less than 80 cm of annual rainfall. Graded bunds may be narrow-based or broad-based. A broad-based graded terrace consists of a wide-low embankment constructed on the lower edge of the channel from which the soil is excavated. The channel is excavated at suitable intervals on a falling contour with a suitable longitudinal grade.

At Dehra Dun, studies on the channel terraces in the alluvial soils of the valley with 3-4 per cent slope have shown that the two vertical intervals (V.I.) tried, viz. 0.3 (S + 2) and 0.3 (S + 3) did not show much difference in run-off, whereas the soil loss per mm of run-off was more from the V.I. = 0.3 (S + 3). However, the annual soil loss was within the permissable limits and, therefore, V.I. of 0.3 (S + 3) can be adopted safely. *V.I. = Vertical interval in metres. S = Slope %

Observations on cross-section revealed that the broad-based and narrow-based cross-sections functioned well without any special maintenance. Broad-based terracing is recommended where farming is practiced with tractors. Studies on broad-based channel terraces constructed on black soils (land slope, 1%) in the semi-arid climate at Bellary showed that a V.I. of 0.6 m was better than that of 0.75 m and that the channel terrace with a variable grade showed less of run-off and soil loss than the channel terrace with a uniform grade. At Kota, graded bunds, 0.7 to 1.0 sq m cross-section and with 0.1 to 0.5% channel grade were found suitable for clay loam to clay black soils.

Bench- terracing

This consists of a series of platforms having suitable vertical drops along contours or on suitably graded lines across the general slope of the land. The vertical drop may vary from 60 to 180 cm, depending upon the slope and soil conditions, as also on the economic width required for easy cultural operations. The material excavated from the upper part of the terrace is used in filling the lower part. A small 'shoulder' bund of about 30 cm in height is also constructed along the outer edge of the terrace.

Bench terraces may be 'table top' or sloping outward or inward with or without a slight longitudinal grade, according to the rainfall of the tract - medium, poor or heavy, and the soil and the subsoil are fairly absorptive or poorly permeable. On steeply sloping and undulating land, intensive farming can be practiced only with bench-terracing. The initial cost of bench -terracing is more than that of bunding. However.

Bench-terracing helps to retain the soil, moisture, manure and fertilizer better and facilitates the application of irrigation, if available. In rainfed areas, terracing is usually practiced on slopes ranging from 6 to 33 per cent. It may have to be used on gentle slopes, if irrigation is to be applied to the crop. The difference in the length of bench-terraces did not influence significantly the run-off, soil loss and the total yield of potato growth on them. However, from the analysis of the fortnightly soil-moisture data, it was observed that on benches longer than 91.5 m, the farther ends of the terraces were significantly drier than those near the cross disposal drains.

On this consideration, it was recommended to have bench terraces, of about 100 m in length in the Nilgiris. Five longitudinal grades of bench terraces constructed on land having 25 per cent slope and with 2.5 per cent inward slope were compared to Ootacamund. The run-off and the soil loss increased significantly with the increase in the grade. However, the maximum run-off was only 0.9 per cent of the total rainfall and the soil loss was only 0.3 tonne/ha under the maximum grade of 0.84 per cent, both of which are negligible. The differences among the yields of the potato crop from bench-terraces of various grades were also not significant. Similarly, the effect of the inward grade of the bench terraces on the yield of potato was not found to be significant.

Grassed Waterways

A grassed waterway is associated with channel terraces for the safe disposal of concentrated run-off, thereby protecting the land against rills and gullies. A waterway is constructed according to a proper design and a vegetative cover is established to protect the channel section against erosion because of the concentrated flow. In the alluvial soils of the Doon Valley under the humid subtropical climate, Panicum repens was the best-suited grass, followed by Brachiara mutica, Cynodon plectostachyus, Cynodon dactylon and Paspalum notatum. The suitability of a grass was based on the cover it gave, the ease with which it was established and the forgae yield obtained from it.

EVALUATION OF SOIL AND WATER CONSERVATION

The beneficial effects of conserving soils moisture by constructing bunds and levelling the land have been extensively demonstrated in the semi-arid alluvial plains of Uttar Pradesh where 35 per cent, 63 per cent and 98 per cent increase in the yields of kharif and rabi crops has been obtained by bunding alone, levelling alone and bunding-cum-lenelling respectively.

Bunding has increased the yields of Setaria, cotton and johar by 18 per cent, 11 per cent and 17 per cent in large-scale field trials in Tamil Nadu. Contour-bunding has not only increased the yield of crops in the dry farming areas in Maharashtra, but has also increased the water level in the wells and the employment oppurtunities as the age of bunding increased.

Estimation of Peak Rate Run-off

For designing various structures and carrying out mechanical measures to control erosion for the safe disposal of run-off, such as grassed waterways and check dams, the estimation of peak rate of run-off is an important design criterion. Three monographs have been developed to estimate the peak rate of run-off by using three different method\s, viz. The Rational Method, the Cooks Method and the Hydrologic Soils-Cover Complex Method.

Gully Erosion Control

It is estimated that about 2.3 m ha (5.75 m acres) of land in India is affected by severe gully erosion. These gullies are an indication of very bad management of the land resources. Apart from the fact that the gullied land has been completely destroyed, the gullies are a menace to the adjoining tablelands. Whereas, as a result of gully erosion, the land resource, one of the most important natural resources we have, is completely lost, the people along the river banks suffer from innumerable hardships.

The ravine lands pose a socio-economic problem as they harbour dacoits and undesirable social elements. Also, the ravine lands are one of the chief sources of sediment which chokes the reservoirs in the country. The gently sloping nature of the land, the loamy-sand, the sandy-loam and the silty-loam texture of the soil, intense rains in the monsoon, the improper land use by resorting to overgrazing, the biotic interference with the natural vegetative cover and the faulty agricultural practices are the chief causes of gully erosion all over India.

Classification of Gullies

Any system of gullies has an independent catchment, with a regular stream which has been termed the "drainage system". In each drainage system, it has been observed that gullies with defined side slopes, bed width and depths occur in a regular order. In the upper reaches of the drainage system, the gullies are wide and shallow, with varying side slopes.

The middle part of the drainage system is usually deeper, wider and has uniform side slopes normally up to about 15 per cent. The lower portion of the drainage system which is nearer to the main river is usually very deep, has steep side slopes and is associated with intricate branch gullies. The gullies have been classified as follows:

G1: Very small gullies	Up to 3 m deep. Bed width not greater than 18 m. Side slopes vary.
G2: Small gullies	Up to 3 m deep. Bed width greater than 18 m. Side slopes vary.
G3: Medium gullies	Depth between 3 and 9 m. Bed width not less than 18 m. Side slopes uniformly sloping betweeen 8 and 15 per cent.

G4: Deep&narrow gullies	(a)3 m to 9 m deep. Bed width less than 18 m. Side slopes vary. (b) Depth greater than 9 m. Bed width varies. Side slopes vary, mostly steep or even vertical; with intricate and active branch gullies.

While the general pattern of land-capability classification is based on the recommendations of the Soils-Survey Manual, the interpretation in relation to slope classes and the susceptibility to erosion hazard by active gullies has been modified to suit the conditions in ravine lands. In ravine lands, the slope of the land and the nearness of gullies are the most important and dominant factors which vary widely and influence land capability.

Control of gullies and their reclamation for various land uses. The practices described below are arranged in the order in which they should be adopted; the cost of their execution increases progressively. Closure to grazing and other biotic interference. A study of the natural flora of the eroded areas has shown that only poor types of annual and unpalatable grasses are growing in place of the desirable climax associations which should normally exist in the prevailing soil-climatic environment.

It is reported that at Vasad in Gujarat, as result of closure to grazing and other biotic interference, (a) Aristida funiculata and Themada triandra were first replaced by Apluda mutica (a tall annual grass) and then by the perennials, such as Eremopogon foveolatus, Heteropogon contortus, Dicanthium annulatum and Cenchrus sp., (b) the loss of soil and water progressively decreased from the area as the natural vegetation improved, and (c) there was not only qualitative but quantitative increase in the yield of grasses. Similar results have been reported on the clay loam and clay soils of the Chambal ravines where closure to grazing and biotic interference resulted in the qualitative and quantitative increase of useful species, such as Dicanthium annulatum, Cenchrus ciliaris and Apluda mutica. Under the upper Damodar catchment conditions, it has been observed that closure to grazing reduced the soil loss from 3.3 tonnes/ha (under overgrazing condition) to only 0.6 tonnes/ha.

Checking the Growth of Gullies

Contour and peripheral bunding

After the closure of the ravine lands, the immediate problem to be tackled is to stop the devastating rate of soil erosion and the growth of gullies and the conversion of the cultivated tablelands into wastelands for this purpose, it is essential to retain as much precipitation a possible on the land in the semi-arid and sub-humid areas and safely dispose of the excess run-off in humid areas. In the semi-arid and the sub-humid area of India, contour and peripheral bunds of various cross-sections have been found ideal for this purpose.

In Gujarat, it is now recommended that for alluvial soils the contour and peripheral bunds should be 0.9 sq m to 1.2 sq m in cross section and should be spaced at 90 cm to 120 cm vertical intervals. The contour and peripheral bunds are stabilized by sodding with Dicanthium annulatum, Cenchrus ciliaris and Panicum antidotale. It has been observed at Vasad in Gujarat that land directly under bunds can give a net annual income of Rs 385 per ha from the sale of these grasses. The excess run-off is safely disposed of through grassed ramps or pipe outlets. In the Chambal ravines, where the soil are relatively fine and the rainfall is heavier, graded bunds in association with grassed waterways and drop structures for the safe disposal of water have been found to be better than the contour lands.

Gully Plugs of Various Materials

The eroding and deepening of gully beds can be prevented with gully plugs. Gully plugs protect the gully beds by reducing the speed of run-off water, redistributing it, increasing its percolation, encouraging silting and improving the soil-moisture regime for establishing a plant cover. Gully plugs of various materials, e.g. brushwood live hedges, earth, sand bags, brick masonry and boulders have been tried in India. The size and material of the gully plugs depends on the width, length, and the bed slope of a gully and the anticipated run-off.

All types of gully plugs are effective either in retaining or retarding the run-off. Earth is the cheapest and most readily available material and it is, therefore, easire and economical to construct the earthen gully plugs, wherever possible. Boulder gully plugs are equally effective if the material is available. Brushwood is often available, but there is likely to be a shortage of wooden posts. Moreover, brushwood gully plugs cannot survive the severe attack of white ants, in spite of the fact that the wooden posts are coated in tar, they disintegrate in a couple of years.

In the long run, they prove to be costly. For the same considerations of short life and high costs, the sand-bag gully plugs are not recommended.

Gully plugs of brick masonry are constructed at the confluence of all gully branches of a compound gully. For gullies where no run-off is expected from the top, earthen gully plugs of 1.1 sq m cross-section (with a grassed ramp, 22.5 cm below the level) and spaced at 45-60 m horizontal intervals are suitable; for gullies in which run-off from the top is expected, an earthen gully plug of 2.2 m cross-section with a pipe outlet is to be provided.

The diameter of the pipe shall be determined by the catchment area (15 cm diameter A.C.C. spun pipe up to a discharge of 2 to 3 cusecs from an effective catchment of 1.6 ha is suitable). A composite check dam of earth and brick masonry (spillway portion) is necessary for larger catchments (more than 1.6 ha). This is located at the confluence of a big compound gully with the main drainage system or in the bed of the main drainage system at 1.2 m vertical intervals or 120 m horizontal intervals.

Reclamation of Gullies

The belief that all the gullied land can be reclaimed for cultivation by making level benches in gully beds is not feasible. This may be possible if no limit is fixed on expenditure. It has been observed that small and medium gullies can be conveniently, safely and economically reclaimed for cultivation, whereas deep and narrow gullies should be retired to a permanent vegetative cover of grasses and trees.

Small Gullies

Small gullies are reclaimed by clearing, minor levelling with bulldozers and constructing diversion-cum-check bunds of 1.5 sq m cross-section spaced at horizontal intervals of 30-45 m. Grassed ramps are provided for the disposal of excess run-off at the ends of bunds near the gully sides. Consecutive grassed ramps are diagonally opposite to one another so that the path of run-off is lengthened and its flow velocity is reduced so as not to cause any scouring. At the end of the small gully, a composite check dam of earth and brick masonry spillway portions constructed.

Medium Gullies

A small gully gets transformed into a medium gully along the length of the main drainage system. A medium gully is reclaimed by clearing and levelling the bed and constructing a series of composite earth and brick masonry check dams at vertical intervals of 1.2 m (which gives a horizontal interval of 120 m on 1% slope of the gully bed) and terracing the side slopes.

Side Slopes Terracing

The uneven side slopes of the medium gullies having 8 to 15% slope are bench-terraced into level terraces at 0.9 to 1.2 m vertical intervals. The terraces are given a back slope of 1 in 50 and a longitudinal grade of 1 in 200 towards the grassed outlet. A ridge bund of 0.3 sq m cross-section is provided at the edge of each terrace. Terrace faces are given a slope of 1.5 : 1. Bench terraces are constructed when the gully sides are having a uniform slope for a length of at least 120 m to justify the cost of terracing.

The terrace faces, grassed outlets and earthen checked dams are stabilized by sodding or growing them with suitable grasses. Dicanthium annulatum and Cenchrus ciliaris have been found suitable for this purpose, specially in Gujarat and in the ravine son the banks of the Jamuna river. Bench terraces and the check dams require careful maintenance for the first two years in view of the unsettled conditions of the soil.

Deep and Narrow Gullies

As mentioned earlier, the best land use of ravine lands is to retire them to permanent vegetation comprising grasses and trees. This type of land use is a must for the deep and narrow gullies. The natural tree species of the ravine lands of Gujarat comprise Acacia nilotica, A. senegal, A. leucophloea, Azadirachta indica, Albizia lebbeck, Feronia elephantum, Prosopis spicigera, etc. With closure to grazing and other biotic interferences, these species make very good growth which is further helped by silvicultural operations, such as climber-cutting and thinning.

Afforestation with a number of species has been tried. Acacia nilotica, Ailanthus excelsa, Albizia lebbeck, Azadirachta indica, Dalbergia sissoo, Dendrocalamus strictus, Eucalyptus camaldulensis, E. citridora, E. hybrid, Pongamia glabra, Phyllanthus embilica, Salmalia malabarica and Tectona grandis are quite promising species for ravines in Gujarat.

For the afforestation of ravines along the banks of the Yamuna river at Agra, Dalbergia sissoo, Acacia nilotica, Azadirachta indica, Albizia lebbeck, Pongamia pinnata, Holoptelia integrifolia, Dendrocalamus strictus and Eucalyptus hybrid have been found to be the most suitable species. In Kota, Rajasthan, Dendrocalamus strictus, Salix tetrasperma in gully beds and Acacia nilotica on the top and marginal lands have been found to be very suitable.

Landslides

Landslides pose an immense threat to highways, villages, agricultural lands, lines of communication, etc. The problem of landslides is more serious in the Himalayan region than anywhere else in India. One landslide about 4 ha in area in Nalotakhala watershed at Dehra Dun has been controlled.

The Nalota Nala landslide was controlled by protecting the landslide against the scouring action of the torrent (by training the lower reaches of the torrent, stabilizing the debris cone, stabilizing the middle reaches of the torrent), by stabilizing the landslips and landslides (by wattling to provide temporary support to bare erodible slopes with grass and brush, by planting deep-rooted shrubs and grasses, such as Pennisetum purpureum, Ipomoea carnea, Vitex negundo and Pueraria hirsute).

Based on the observations on the landslide control, it is recommended that the control methods may be adopted, depending on the kinds of materials available in the area, the cause of the problem, the type and the rate of the movement of the slide. These factors vary so greatly from slide to slide that details of remedial measures are required to be tailored to deal with each slide.

References

Bonsu, M., "Organic residues for less erosion and more grain in Ghana," In: Soil Erosion and Conservation, El-Swaify, Moldenhauer and Lo (eds)., *Soil Cons. Soc. Amer.*, Ankeny, Iowa. 1985.

Stewart B.A., Unger P.W. and Jones O.R., "Soil and water conservation in semi-arid regions," In: Soil Erosion and Conservation, Ei-Swaify S.A., Moldenhauer W.C. and Lo A. (eds), *Soil Cons. Soc. Amer.*, Ankeny, Iowa, 1985.

Walker P.J., "Cropping and soil conservation in semi-arid western New South Wales (Australia)," *J. Soil Conservation Service of New South Wales* 38(2): 49-56, July 1982.

Willcocks T.J., "Tillage requirements in relation to soil type in semi-arid rainfed agriculture," *J. Agric. Eng. Res.*, 30: 327-336, 1984.

5

Irrigation Water Conservation

Irigation is the artificial application of water to the land or soil. It is used to assist in the growing of agricultural crops, maintenance of landscapes, and revegetation of disturbed soils in dry areas and during periods of inadequate rainfall. Additionally, irrigation also has a few other uses in crop production, which include protecting plants against frost, suppressing weed growing in grain fields and helping in preventing soil consolidation. In contrast, agriculture that relies only on direct rainfall is referred to as rain-fed or dryland farming. Irrigation systems are also used for dust suppression, disposal of sewage, and in mining. Irrigation is often studied together with drainage, which is the natural or artificial removal of surface and sub-surface water from a given area.

In the middle of the 20th century, the advent of diesel and electric motors led for the first time to systems that could pump groundwater out of major aquifers faster than it was recharged. This can lead to permanent loss of aquifer capacity, decreased water quality, ground subsidence, and other problems. The future of food production in such areas as the North China Plain, the Punjab, and the Great Plains of the US is threatened.

At the global scale, 2,788,000 km^2 (689 million acres) of agricultural land was equipped with irrigation infrastructure around the year 2000. About 68% of the area equipped for irrigation is located in Asia, 17% in America, 9% in Europe, 5% in Africa and 1% in Oceania. The largest contiguous areas of high irrigation density are found in North India and Pakistan along the rivers Ganges and Indus, in the Hai He, Huang He and Yangtze basins in China, along the Nile river in Egypt and Sudan, in the Mississippi-Missouri river basin and in parts of California. Smaller irrigation areas are spread across almost all populated parts of the world. Only 8 years later in 2008, the scale of irrigated land increased to an estimated total of 3,245,566 km^2, what is nearly the size of India.

Sources of Irrigation Water

Sources of irrigation water can be groundwater extracted from springs or by using wells, surface water withdrawn from rivers, lakes or reservoirs or non-conventional sources like treated wastewater, desalinated water or drainage water. A special form of irrigation using surface water is spate irrigation, also called floodwater harvesting. In case of a flood (spate) water is diverted to normally dry river beds (wadis) using a network of dams, gates and channels and spread over large areas. The moisture stored in the soil will be used thereafter to grow crops. Spate irrigation areas are in particular located in semi-arid or arid, mountainous regions. While floodwater harvesting belongs to the accepted irrigation methods, rainwater harvesting is usually not considered as a form of irrigation. Rainwater harvesting is the collection of runoff water from roofs or unused land and the concentration of this. Some of Ancient India's water systems were pulled by oxen.

Around 90% of wastewater produced globally remains untreated, causing widespread water pollution, especially in low-income countries. Increasingly, agriculture is using untreated wastewater as a source of irrigation water. Cities provide lucrative markets for fresh produce, so are attractive to farmers. However, because agriculture has to compete for increasingly scarce water resources with industry and municipal users, there is often no alternative for farmers but to use water polluted with urban waste, including sewage, directly to water their crops. There can be significant health hazards related to using water loaded with pathogens in this way, especially if people eat raw vegetables that have been irrigated with the polluted water. The International Water Management Institute has worked in India, Pakistan, Vietnam, Ghana, Ethiopia, Mexico and other countries on various projects aimed at assessing and reducing risks of wastewater irrigation. They advocate a 'multiple-barrier' approach to wastewater use, where farmers are encouraged to adopt various risk-reducing behaviours.

These include ceasing irrigation a few days before harvesting to allow pathogens to die off in the sunlight, applying water carefully so it does not contaminate leaves likely to be eaten raw, cleaning vegetables with disinfectant or allowing fecal sludge used in farming to dry before being used as a human manure. The World Health Organization has developed guidelines for safe water use.

Fifty years ago (since 2010), the common perception was that water was an infinite resource. At that time, there were fewer than half the current number of people on the planet. People were not as wealthy as today, consumed fewer calories and ate less meat, so less water was needed to produce their food. They required a third of the volume of water we presently take from rivers. Today, the competition for water resources is much more intense. This is because there are now more than seven billion people on the planet, their consumption of water-thirsty meat and vegetables is rising, and there is increasing competition for water from industry, urbanisation and biofuel crops. To avoid a global water crisis, farmers will have to strive to increase productivity to meet growing demands for food, while industry and cities find ways to use water more efficiently.

Successful agriculture is dependent upon farmers having sufficient access to water. However, water scarcity is already a critical constraint to farming in many parts of the world. With regards to agriculture, the World Bank targets food production and water management as an increasingly global issue that is fostering a growing debate.

Physical water scarcity is where there is not enough water to meet all demands, including that needed for ecosystems to function effectively. Arid regions frequently suffer from physical water scarcity. It also occurs where water seems abundant but where resources are over-committed. This can happen where there is overdevelopment of hydraulic infrastructure, usually for irrigation. Symptoms of physical water scarcity include environmental degradation and declining groundwater.

Economic scarcity, meanwhile, is caused by a lack of investment in water or insufficient human capacity to satisfy the demand for water. Symptoms of economic water scarcity include a lack of infrastructure, with people often having to fetch water from rivers for domestic and agricultural uses. Some 2.8 billion people currently live in water-scarce areas.

Irrigation Methods

Sources and quality of irrigation water vary between piped water and raw wastewater. Most common is, however, the use of stream and drain water, highly polluted with domestic grey water. A brief overview of some key features in Accra, Kumasi and Tamale is presented here:

In Accra, the main source of irrigation water is urban drains. The contents of these drains vary from raw wastewater as in Korle-Bu to storm water diluted wastewater as in Marine Drive, though this changes with seasons. In Dzorwulu, a polluted stream (Onyasia) is used in combination with pipe-borne water. Other than using a big drain that runs through Accra's La area, a few farmers there also use partially 'treated' wastewater' from the maturation pond of the stabilisation pond treatment system belonging to the Burma military camp. Other farmers in La use piped water.

In Kumasi, polluted rivers and streams are the main sources of water for 70 % of the farmers. None of the farmers interviewed used (raw) effluent directly from the source or a sewage treatment plant. A very few cases (n< 5) were recorded where farmers, because they have no choice, use wastewater from drains. There is an extensive use of shallow dug wells on valley bottoms (27%), especially in the urban area.

Of the 70 farmers interviewed, more than 75% said that they use the source of water that is accessible and reliable. Piped water is not only expensive but is unreliable and in any case inaccessible to most farmers. In Tamale, with no perennial stream and a long dry season, water is scarce. Some farmers end up using drain water. In areas like Kamina, farmers use wastewater from a broken sewage treatment plant and others at "Waterworks" use water from a water supply dam, which has been abandoned due to water pollution. The agricultural use of faecal sludge in Tamale does not concern vegetables and targets its nutrient value, not the water.

Watering cans, buckets, motorised pumps with hosepipe, surface and sprinkler irrigation methods, as described below, are being used in the study areas.

Watering Cans

This is the most common irrigation method used in all the study areas. It is also the most precise one for fragile leafy vegetables. Farmers use watering cans to fetch and manually carry water from a water source, mostly shallow dug wells, streams or dugouts, to the fields, followed by watering of crops through the spout or shower head of the can making it an overhead irrigation method. In many cases, farmers carry two watering cans at a time.

As men dominate irrigated urban farming, it is rare to see a woman with two watering cans. In peri-urban areas, where women are more common, they seem to prefer water fetching and application with buckets, often transported as head load. One watering can as used in Ghana has a capacity of 15 litres of water. Almost all farmers in the valley bottom of urban Kumasi use watering cans.

Most of them have shallow dug wells on their farms and even for those who have to fetch water from streams, the distance is usually short (10-15 m). Previous studies showed that farmers closer to water sources tend to over-irrigate in absolute crop water requirements. However, farmers are trying to keep their leafy vegetables fresh, thus irrigation is just wetting the soil surface and evaporation losses are significant.

Bucket Method

In this method, bowls and buckets are used to fetch water, usually from a stream/river or dugout. It is then manually carried to the fields where it is either applied directly or put in a drum to be applied later. This practice mostly involves women and children carrying buckets as 'head loads' and is commonly done in the peri-urban areas. Here male farmers can easily involve family members and take advantage of the traditional role of women and children in transporting water.

Farms are comparatively further from the water source than the ones where watering cans are used, but normally are less than 50 m. The manner of watering is determined by crop height and type. Farmers using buckets and watering cans come in contact with water mainly by stepping in it while fetching, or water splashing on them while carrying and during watering. Crop contamination is very high due to the combination of crops with large surface area and overhead application.

Motorised Pumps

Motorised pumps are mostly seen in peri-urban areas, but also increasingly in Accra. A small motor pump is placed temporarily near a water source, usually the bank of a river or a big stream and water is pumped through rigid plastic pipes or semi-flexible pipes which are connected to a flexible hosepipe at the end. Farmers use the hose to apply water to their crops either overhead or near the roots on the surface. In other cases, pumps helped to reduce transport ways: water was pumped into a dugout from where water was fetched with cans.

In many cases where motorised pumps were used in Ghana, there were massive water losses with many pipes of inappropriate size or leaking. Farmers finally end up flooding fields along the pipelines and at the point of irrigation. More than one farmer is needed for the operation. Irrigators are often fully wet as they try to fix the pump, pipes and direct the hose for irrigation. The fields are usually adjacent to the water sources and the pipes could be as long as 300 m.

Due to the high velocity of water from the pipes, watering can be done overhead even for tall crops like mature garden eggs. As the water pressure and the hose would damage leafy vegetables, usually only taller growing and stronger vegetables are irrigated in this way. Though the total amount of water applied per season using this method was high (5 litres/second), the distribution and uniformity of application was poor. Pumps are hired from the 'wealthier' farmers for a constant fee per day.

In order to make maximum use of their money on the day of hire, farmers end up over-irrigating the areas most accessible to the feeder hose, leaving other areas of the field under-irrigated. In Dedesdua, a village in peri-urban Kumasi, to reduce costs, farmers on average irrigated once in three weeks (for a 120 day crop growth period) instead of once a week. This was equivalent to 5-10 hires over the 120-day period instead of 20 hires, which is a considerable saving for the farmer. They felt that this was sufficient for the crops but in reality such long cycles could affect crop productivity.

Form of Surface Irrigation

Some form of surface irrigation, mainly furrow is being practiced in La in Accra. La farming area is a comparatively wider open space with a topography that allows for furrow irrigation. The source of water is a drain that runs from the nearby military camp to its treatment plant. Farmers have constructed an open weir and diversion channel to irrigate their plots downstream by furrows, or they divert water into dugouts from where they can fetch with a watering can. During the dry season, farmers raise the water level in the drain with sand bags and divert the water in a main canal, which conveys the water to the plots. Furrow irrigation can reduce crop contamination since crops are grown on ridges, but exposure to farmers is as high as with water fetching from streams and drains.

Sprinkler Irrigation

This method was seen in a few sites (behind Georgia Hotel in urban Kumasi and Dzorwulu in urban Accra). In both cases, the sprinkler system is connected to a pipe borne water source. Low cost materials were used, like bamboos as sprinkler risers etc. These systems are the portable type and farmers in Dzorwulu combine them with the watering can method. The fields were reasonably large but the crops grown were the same as in the other areas. In this case, irrigation water has both on and off farm effects (aerosolised particles) on crops, farmers and the environment.

Land Productivity

Leafy vegetables, which are the most commonly grown crops in irrigated urban agriculture, have higher and more regular crop water requirements compared to more traditional crops. According to Agodzo et al. irrigation water requirements of most vegetables grown in Ghana vary between 300 and 700 mm depending on the climatic conditions and crop species. As the extension service has limited training to support informal irrigation, farmers have learnt over time when and how much water to apply to their crops.

Generally, most farmers irrigate in the mornings and evenings, saying that at these times "it is cooler so we can more easily carry the water-load" which corresponds well with periods of low evapotranspiration rates, allowing other jobs during normal working hours (8am to 5pm). Not all farmers can afford to buy irrigation equipment, like motorized or electrical pumps. However, neighbourhood arrangements enable farmers to hire pumps on affordable terms.

At Dedesua, in Kumasi, for instance, most farmers only pay for the fuel of a local motor pump. Payment can also be made on flexible terms such as paying after selling the crop or by providing labour for the pump owner. Some farming sites have farmers associations to exchange labour and irrigation equipment. From field monitoring of water use, some farming sites in and around Kumasi showed a tendency towards over-irrigation in the urban areas by one-third of the irrigation water requirements and under-irrigation of the same magnitude in the peri-urban areas.

Urban farms are much smaller than peri-urban farms and farmers predominantly use watering cans for irrigation. Urban farmers achieve a more uniform spatial water distribution, though, because of the watering cans, and their irrigation intervals, which are regular. Peri-urban farmers use either buckets or motor pumps connected to water hoses. Periurban farmers have irregular irrigation intervals and poorer water distribution, especially those using hosepipes. As only a few peri-urban farmers own a pump, they wait for long periods before irrigating their farms 'queuing to hire a pump'.

Subsequently it is quite common among these farmers to apply as much water as they can when the pump is available. The bucket method is laborious and depends on availability of women and children. This too contributes to irregular watering. A good overview on tomato, garden egg, pepper, okra and onion production and yields in Ghana was provided by Nurah, but few studies have been conducted on the productivity of vegetables common in Ghana's cities.

Indicative studies on urban lettuce and cabbage production show typical production levels ranging between 20-35 tons/ha (lettuce, fresh weight) and about 40 tons/ha (cabbage) per crop. With e.g. 5 lettuce crops and 3 cabbage cops per year, the annual output is significant. Some variations across cities and seasons have been observed. For example, production levels in Kumasi are generally higher (10-20%) than in Accra, probably due to better soils except under excess rainfall, when levels start declining in Kumasi's inland valley bottomlands.

Production during the wet season is higher in both cities than during the dry season. Preliminary calculations of water productivity under the Challenge Program on Water & Food project CP38 show a typical range of 6-9 kg/m^3 for lettuce grown in Kumasi, and about 8 kg/m3 for tomatoes. However, the lettuce values could also be lower as calculations did not consider the period when seedlings are in the nursery bed.

Constraints to Technology Change

Having to spend a significant share of time on irrigation farmers desire less laborious and cheap irrigation methods that can reduce their workload. However the continued use of arduous methods of irrigation in urban agriculture, even when newer technologies are available, raises the question of the reason why these are not more widely used.

In Accra, farms are often found along streams and drains, and are at best tolerated by the authorities. The transport distance for the watering can is usually short enough to favour labour over capital input. Watering cans allow more flexibility, one-man-usage, and are less sensitive to bad water quality and solids. Moreover, they allow "soft" water application protecting young vegetables on their beds. All these are good reasons to avoid investment in pumps and hoses. In addition, there are differences in the related input markets between Francophone Togo and Anglophone Ghana as well as in their practices and their promotion. Enterprise Works, for example, started promotion of treadle pumps in most Francophone countries of West Africa between 1995 and 1999, while corresponding activities in Ghana only started in 2002. A comparison with neighbouring Lomé, where farmers cultivate high value crops for export on poor quality beach sands using motorised pumps, shows that there is a combination of factors involved, which goes beyond the higher investment and maintenance costs of such technologies.

Farmers in Lomé have access to larger plots, and the city authorities accept them. In many cases, tenure agreements exist. This security favours investments, for example, in tube wells and multiple storage ponds, i.e. transport saving technologies, which are necessary to maximise the profits from larger plots. Thus it becomes again obvious that technology promotion has to consider a variety of local conditions, both biophysical and socioeconomic. Shallow wells and watering cans may be the most appropriate technology, for example, in Kumasi's inland valleys.

The demand for treadle or motor pumps might, however, rise on upland sites, and in Accra, especially where farmers can share one pump and the water sources are not too close. Treadle pumps might be tried where groundwater is available between 1 and 7 meters or the walking distance between field and water source is more than 50 m. If the pump is not mobile, the farm site needs facilities to secure the pump overnight as farmers might live far from their fields. Wherever there is better water quality than in drains, low-cost drip irrigation technologies like bucketkits and drum-kits, could be tested.

Faecal Sludge

Farmers in Tamale and other northern parts of Ghana use faecal sludge in agriculture. Sludge is disposed of on farms during the dry season, allowed to decompose and mixed with the soil at the start of the rainy (farming) season. Generally, men are responsible for the acquisition (from septic trucks) and application of faecal sludge on farms. This is because men generally own the land due to the traditional setup as there are no women fields.

Cofie et al. conducted interviews with 100 farmers using faecal sludge in Tamale and Bolgatanga (another urban center in northern Ghana). Half of the farmers were using tractors for land preparation, while the rest used hoes and bullocks. The average farming area is about one ha and though some farmers claimed to have been using faecal sludge for more than 25 years (Gumani community in Tamale), an average of about five years of use was recorded.

Many farmers own land with less than 20% farming on family or hired land. About 63% received faecal sludge from the local assemblies' tankers, though they could wait for 1-3 weeks before delivery after making an unofficial request. Four out of five farmers said that they used faecal sludge to increase crop yields and improve fertility, however up to half of them experienced social constraints in its use.

Methods of Faecal Sludge Application

Methods of sludge application have been derived from the experience and understanding of the farmers. Over the years, farmers in Tamale have taken advantage of the climatic conditions to develop a safe method of faecal sludge application in their fields. The high temperatures lead to effective drying of the discharged sludge and allow the sludge to be handled easily while integrating it into the soil.

Moreover the health risks associated with use of faecal sludge are reduced, as most microorganisms contained in the sludge die in high temperatures. By the start of the first seasonal rains, most of the sludge is completely dry and ready for use. Most farmers grow cereals on the fields fertilised with faecal sludge, i.e. not vegetables. Two main methods are used:

— *Surface spreading*: This involves discharging faecal sludge at various points (accessible to the septic emptier) at random on farmers' plots. This is done during the dry season (October -December). By the end of the dry season (February-March), the faecal sludge applied becomes very dry. Farmers then gather and redistribute this material evenly on the field, before cultivation.

— *Pit method*: Pits are dug on farms and rice and maize straw is placed at the bottom of the pit. Faecal sludge is then poured into the pit, which is large in size and can take several trips of the conveying truck. Layers of bran and straw are placed in between subsequent trips. The process is repeated until the pit is full. This is left to compost for months. Before the cropping season starts, the pit is emptied and the dry mixture of faecal sludge and straw is applied evenly on the field. The pit method is not as widely used as the surface spreading method because it requires quite high quantities of crop residues in combination with the faecal sludge.

Nutrients Supplied Through Sludge Application

From farmers' experience over the years, five trips of the suction truck of 4.5m^3 are used to fertilise one acre of land (0.4 ha). Farmers apply an average of 56 m^3/ha. Through this practice, significant amounts of plant nutrients in terms of nitrogen (N), phosphorus (P) and potassium (K) are returned to the soil, and in addition the organic matter level is gradually built up. Based on the average concentration of nutrients in human excreta as reported by Drangert, estimated amounts of N, P, K and carbon in the applied sludge are presented in Table 1.

Table 1: Estimated amount of nutrients applied in faecal sludge by Tamale farmers

Nutrient	Total in human faeces (kg)	Total in m^3 (kg/ha)	Amount applied
Nitrogen (as N)	4.5	8.2	459
Phosphorus (as P)	0.6	1.1	61
Potassium (as K)	1.2	2.2	121
Carbon (as C)	11.7	21.3	1183

Source: Calculated after Drangert based on the nutrient concentration in human feaces per person-year. Density for faecal sludge is taken as 0.55 kg/m^3

This estimate does not consider loss during sludge storage in septic tanks and the amount lost in the field beyond the reach of plants. Based on the current sludge application rate by farmers, about 550 hectares of land can be fertilised annually using the faecal sludge that is generated in Tamale municipality alone at the current collection rates.

In fact, because there is no sanitary sludge storage facility near Tamale, trucks dump the sludge in natural depressions around the city. The new landfill site with sludge sedimentation ponds will not benefit farmers and reduce a viable option for resource recovery towards a closed nutrient loop.

Constraints for Faecal Sludge Use

Among the same set of farmers interviewed by Cofie et al., 74% expressed that they had no problem with using faecal sludge in agriculture. Those who noticed problems complained of foul smell and health issues like itching and foot rot, with very few saying that it attracted public mockery. The major constraints are highlighted below. The interviews were carried out before the new treatment ponds near Gbalahi were put in place:

— Bad odor is a deterrent to its application on farms. Farmers are not allowed to use sludge in the city because of the odor.

— Farmers who have fields near large concentration of houses are unable to use faecal sludge to improve soil fertility. Faecal sludge use is more in peri-urban areas.

— Negative attitude of other people towards the use of human waste. For farmers on hired land, it is a common situation that landowners do not allow them to use sludge on their lands despite its positive effects on soil fertility. Some people also shun the consumption of crops cultivated with faecal sludge.

— There is excessive weed infestation after the application of faecal sludge.

— Land for farming is scarce around the Tamale Municipality.

— Inability of the suction truck drivers to send faecal sludge to fields that are too distant from the city.

Like in Accra's faecal sludge treatment plant in Teshie-Nungua, Tamale's new sanitary landfill site at Gbalahi has the provision to use settled sludge to enrich composted solid waste. However, besides differences in quality, this would require that farmers organise and pay for compost transport.

The current system of sludge dumping on farmers' fields appears more like a win-win situation as long as the environmental and health risks are under control. An alternative form of cocomposting is under test in a Buobai near Kumasi. Due to the availability of cheap poultry manure, farmers' demand is, however, low.

Surface Irrigation System

To achieve high efficiency and uniformity of irrigation using unpressurised gravity water application methods, all parts of an irrigated field should receive water for near equal lengths of time, with a minimum of water lost to runoff or to deep percolation below the root zone. Since the soil is quite porous, the flow rate of water is continually reduced as it spreads. Thus, getting water uniformly spread over a field requires careful management. There must be a balance between the size and shape of a field, slope, water flow rate, sod infiltration characteristics, and surface roughness, such as soil clods or vegetation that retard water flow.

Soils with high infiltration rates are not wellsuited for surface irrigation. Either field sizes must be kept small, or large flow rates are needed to achieve even applications. Soils with moderate to low infiltration rates (loams and other finetextured soils) are better for surface irrigation. Fields should have a uniform grade in the direction of water flow to obtain best results. A uniform grade or level surface prepared through precision land levelling helps to provide equal infiltration opportunity times.

Field slopes in excess of 3% are not recommended, unless the field is planted to a permanent sod-forming crop or unless other means of erosion control are used. Very low flow rates are used in this situation. The water flow rate into a field or furrow is an important design factor. It must be carefully balanced against the soil type and slope so that erosion is minimised, and against the field slope and length so that the water reaches the end in a reasonable amount of time. Operating the system at flow rates below or above the design flow rates can lead to inefficient and nonuniform applications.

Once the water reaches the end of the field, growers must manage the inflow to reduce tailwater runoff losses. There should be some means of controlling the direction of water flow. In furrow or rill irrigation the furrows act as conveyance channels. The size, shape and spacing of the irrigated furrows affect the soil volume wetted. When border irrigation or other flooding methods are used, dikes or borders are constructed to contain the flow within the desired field area.

Management of Surface Irrigation

Surface irrigation systems can be as efficient as most other methods. This requires improving the management and control of water, knowing how much water is applied and scheduling applications according to soil water levels and crop needs. Information on soil moisture monitoring and crop evapotranspiration from Washington's Public Agricultural Weather Stations (PAWS) and Washington Irrigation Scheduling Expert (WISE) are available on the Scientific Irrigation Scheduling (SIS). Typical water application efficiencies for unimproved surface irrigation systems range from 45% to 60%.

Using careful management, improved water control, and reuse of tailwater runoff, growers can boost efficiencies to 70% to 85%. Differences in the amount of water which infiltrates between the top and bottom ends of the field are inevitable. Tailwater runoff is also inevitable if the bottom ends of fields are to be adequately irrigated. Using rill systems and improved management, which is enhanced by improved on-farm water delivery facilities, growers can reduce overwatering losses at the top ends of fields and the runoff losses at the bottom.

Growers will obtain the best water application from the top end of a field to the bottom if the water reaches the end of the field within onequarter of the planned irrigation duration. Where possible, adjust inflow stream size to meet that schedule for the field length. For instance, if you plan to irrigate with 48-hour sets, water should reach the end of the field, or advance, within 12 hours for best results. Use the largest, nonerosive stream size possible to achieve the desired advance time. For most conditions, the maximum safe stream size in gallons per minute per rill is found by dividing the field slope in percent into 10 (10/S,%).

As an example, a field with a 2% slope should have a maximum stream size per rill of 5 gallons per minute. If water cannot be made to advance to the end of the field within the desired time using the maximum stream, then the field is probably too long for that given combination of slope and soil type. Cut the field in half by adding water distribution facilities at mid-length. Tailwater from the upper half can be used to supplement the irrigation of the lower half or caught in a ditch for reuse elsewhere on the farm.

Furrow lengths up to ½ mile on relatively flat slopes (less than 0.5%) can be efficiently irrigated using large streams. As a practical limit, however, field lengths of ¼ mile or less are recommended for most Washington conditions. Surge flow surface irrigation, which has received recent research attention, has provided significant increases in the uniformity of water application from the top to the bottom end of a field. Using surge flow, researchers have been able to advance water to the end of the field at the same rate as with conventional furrow irrigation while using only 50% to 60% of the water.

Longer fields can be more efficiently irrigated with surge flow. Application efficiency of surge flow with tailwater reuse can be as high as 75% to 90%. Once water reaches the end of the field, cut back the stream size by 1/3 to ½ so only a small, but continuous stream of runoff water exits out the end of the rill. This will minimise runoff losses during the remainder of the irrigation. The cutback in flow can be achieved by:

— adjusting the siphon tube setting or using two siphon tubes during the advance phase and removing one after the water reaches the end of the furrow;
— closing the gates part way with gated pipe; and
— closing the valves part way with buried pipelines having risers and adjustable valves.

Vegetation in furrows slows water movement along furrows and rills. Weed control in the bottoms of the furrows is very important for achieving uniform water application and reducing deep percolation losses. Cultivation and the formation of a loose soil surface with soil clods impede water flow and increase water intake. In a water-short year, control weeds chemically and avoid cultivating if at all possible. Otherwise, use a furrow packer or smoother after the cultivator to reduce deep percolation losses. Also applying Polyacrylamide (PAM) after soil cultivation will increase lateral water penetration and decrease soil erosion. PAM should also allow increased furrow flow rates that will improve uniformity of irrigation.

While water is available early in the irrigation season, irrigate in all your furrows or rills to completely fill the soil profile to field capacity and maximise soil water storage. Be careful not to overirrigate, as crop root damage may result. When water becomes short, it will help to irrigate in only one furrow per row on wide spaced plantings, such as orchards, vineyards, and hopyards. This will minimise evaporation losses and reduce water use by the cover crop or weeds.

Use the same furrow all season. For closer spaced plantings, such as asparagus and mint, irrigate in every other furrow and do not switch back and forth. Irrigating in the same furrows all season will help save water by minimising the wetted soil evaporation, reducing the germination of weed seeds and reducing the losses from increased soil intake rates.

Tailwater Runoff

The water which leaves any surface irrigated field as runoff is a loss to that field and to the farm unless there are facilities for catching the water and reusing it on the farm. The amount of runoff is typically between 30% and 50% of the water introduced at the top of the field. If this water is reused it can represent a significant supplemental supply compared with allowing the runoff to leave the farm.

Generally, you can use a common collector ditch across the bottom of the field to channel the water into the headland facilities of a downslope field, or into a small pond for storage and later use. Tailwater ponds are often fitted with low lift pumps to move the water to other areas of the farm or back to the top end of the same field. These ponds can also catch sediment in the runoff water.

Headland Facilities

Keep earthen ditches weed-free. Seal cracks and gopher holes. Silt which accumulates offers natural scaling. Leave this silt in place until the ditch capacity is too greatly reduced. Eliminate seepage losses from earth ditches by using plastic linings or replacement with a concrete-lined ditch or pipeline. Gated pipe and buried low pressure pipelines with risers and valves minimise evaporation losses and weed/ grass problems. Increased water control offered by these systems, particularly the flow rate into each furrow, is an important advantage in improving water management.

Knowing how much water is applied with any method of irrigation is a key to good management. It is more difficult to determine amounts of water applied when using surface irrigation than when using any other method. However, tools and techniques are available, ranging from farm delivery gate measurements to siphon tube measurements, to individual furrow flow measurements. A WSU Extension publication, C0912, Determining the Gross Amount of Water Applied- Surface Irrigation, gives information on converting farm delivery weir flow measurements to gross depth of water applied.

Measuring Water Flow

Measuring water in surface irrigation systems is critical for peak efficiency management. Without knowing the amount of water being applied, it is difficult to make decisions on when to stop irrigating or when to irrigate next. A good irrigation manager should know the flow rate of the irrigation water, the total time of the irrigation event and the acreage irrigated. From this, the total amount of water applied can be determined, which will help determine whether the irrigation was adequate and when the next irrigation should be. Irrigation management decisions should be made based on the amount of water applied and how this relates to the consumptive use demands of the plants and the soil water holding capacity.

Units of Measuring Water

There are many ways to express water volume and flow. The volume of water applied is usually expressed in acre-inches or acre-feet for row crops or gallons per tree in orchards. Flow rate terminology is even more varied. Flow rate is expressed as cfs (cubic feet per second), gpm (gallons per minute) and in some areas, miner's-inches. Below is a description of each.

Acre-inch (ac-in.): An acre-inch is the volume of water required to cover an acre of land with one inch of water. One acre-inch equals about 3,630 cubic feet or 27,154 gallons.

Acre-foot (ac-ft): An acre-foot is the volume of water required to cover an acre of land with 1 foot of water. One acre-foot equals about 43,560 cubic feet, 325,848 gallons or 12 acre-inches.

Cubic feet per second (cfs): One cubic foot per second is equivalent to a stream of water in a ditch 1-foot wide and 1-foot deep flowing at a velocity of 1 foot per second. It is also equal to 450 gallons per minute, or 40 miner's-inches.

Gallons per minute (gpm): Gallons per minute is a measurement of the amount of water being pumped, or flowing within a ditch or coming out of a pipeline in one minute.

Miner's inches: Miner's-inches was a term founded in the old mining days. It is just another way of expressing flow. Some areas in the West still use this measurement unit. Caution needs to be taken because there are Arizona miner's-inches, California miner's-inches and probably some that are locally used. Approximately 40 Arizona miner's- inches equals 1 cfs or 450 gpm.

Pressure or Head (H): People often use the phrase "head of water." A foot of head usually implies that the water level is one foot above some measuring point. However, head can also mean pressure. For example, as the level of water rises in a barrel, the pressure at the bottom of the barrel increases. One foot of water exerts 0.43 pounds per square inch (psi) at the bottom of the barrel. Approximately 2.31 feet of water equals 1 psi. Thus, if a tank of water were to be raised 23.1 feet (2.31 x 10) in the air with a hose connected to it, the pressure in the hose at the ground would be about 10 psi.

Area: The cross sectional area of a ditch is often required to calculate flow. Some ditches are trapezoids and others or more like ellipses. To find the area of a trapezoid, measure the width of the bottom (b) and the width of the ditch at the water surface (s) and add them together. Divide that number by 2 and then multiply by the height (h) of the water. If the ditch is more elliptical in shape, take the depth of the water (h), multiply it by the width of the ditch at the surface (s), divide by 4 and then multiply by PI (3.14). To calculate the cross-sectional area of a pipe, the formula is PI x r^2, where PI is 3.14 and "r" is the radius of the pipe.

Measuring Water Flow in Ditches

Float Method

This method is useful to get a rough estimate of flow. First, choose a 100-foot section of ditch that is fairly uniform in depth and width. Mark the zero point and the 100 ft point with a flag or stick. The 100 ft mark should be downstream from the zero point. For most people, one good, long stride equals three feet. If there is no tape measure available, step off about 33 paces. Next, calculate the ditch cross sectional area. Use an average of several measurements along the ditch.

Now, take a float and place it a few feet up stream from the zero point, in the center of ditch. Once the float hits the zero point, mark the time (probably to the nearest second). Then, mark the time the float passes the 100 ft mark. Record the time. Do this several times. Try to place the float in the center of the ditch flow so that it won't bounce off the sides or get caught up in any weeds. After 5-10 tries, average the recorded times.

The flow rate is determined by calculating the velocity of the water and multiplying it by the cross sectional area of the ditch. First, take the length of the ditch (100 ft) and divide it by the time (in seconds). This will give the surface velocity (speed) in feet per second. However, water at the surface flows faster than water in the center of the flow and it is the average flow or center flow that is needed. Therefore, a conversion factor must be used to determine the mean channel velocity. The factor by which the surface velocity should be multiplied by is a function of the depth of the water in the ditch. Table 2 gives the coefficients to be used.

Table 2. Coefficients to correct surface float velocities to mean channel velocities.

Average Depth (ft)	Coefficient
1	0.66
2	0.68
3	0.70
4	0.72
5	0.74
6	0.76
9	0.77
12	0.78
15	0.79
20	0.80

Find the depth measured on the left and the corresponding coefficient on the right. Then multiply the surface float velocity by the coefficient to obtain the mean channel velocity. Finally, take the cross sectional area of the ditch (ft^2) and multiply it by the corrected velocity (ft/sec) and this will compute the flow rate in cubic feet per second (cfs). To convert to gallons per minute, multiply the cfs by 450.

Tracer Method

This method is very similar to the float method but with one exception, a colored dye or salt is used instead of a float. Estimates of the ditch area are still required. Pour the dye upstream of the zero point, and record how long it takes the dye to travel from the zero point to the 100 ft mark. Then the calculations are exactly the same as the float method. This method often works well if the float keeps getting caught on the sides of the ditch. However, in many cases the dye is difficult to see because of the color of the water itself. Test the dye first to make sure it can be seen. The correction factors used with the float method (Table 2) are not required for the tracer method.

Velocity Head Rod

The velocity head rod is used to measure the velocity of water in a ditch and is relatively inexpensive and fairly accurate. The rod is in actuality a ruler used to measure the depth of the water. The water height is first measured with the sharp edge of the ruler parallel with the flow and the again with the ruler turned 90 degrees. Multiply the velocity by the cross sectional area of the ditch to get cubic feet per second. The velocity head rod method works only for velocities greater than 1.5 ft/sec and less than about 10 ft/sec.

The procedure is:

— Place the rod with the sharp edge upstream. Record the depth of the water (normal depth).

— Place the rod sideways. This will cause some turbulence and the water level will "jump" causing the water level to rise. Record the level again (turbulent depth).

— Subtract the normal depth from the turbulent depth and this will be the jump height.

— Find the corresponding velocity.

— Multiply the velocity by the cross sectional area of the ditch to get the flow rate (cfs).

Types of Weirs

There are several different types of weirs that can be constructed and used to determine the flow rate in a ditch or stream. The three most common weirs are:

— V-Notch or Triangular

— Rectangular and

— Cipolletti.

The simplest design is to make the weir out of a sheet of plywood or sheet metal. Cut the wood or metal to fit ditch with the particular shape notch cut out of the top. Make sure the weir is sturdy enough to hold up against the flow of the water. The top two are rectangular weirs. The first is a rectangular contracted weir and is one of the most commonly used. The second is another rectangular weir but since the sides of the weir are actually the sides of the ditch, it is called a suppressed rectangular weir. This type of weir has a trapezoidal shaped notch. The last type shown is a triangular or Vnotched type. With proper installation, all of these weirs can be accurate.

An estimate of the actual flow rate must be made before construction of the weir in order to make sure the notch size is correct. For the V-notch, the dimension requirements are the same and for the Cipolletti, the requirements are also the same but with a 25% slope rising outward at the sides of the notch. To measure the head or height of the water for these weirs, pound in a stake about 6 feet upstream so that the top of the stake is even with the bottom of the notch in the weir. Once in place, the water will rise behind the weir. Measure the depth of water above the stake. The length (L) refers to the width of the opening at the base of the weir notch.

Caution

Installing a weir in a ditch will cause the water behind the weir to rise. Make sure there is enough freeboard or the water in the ditch will overflow.

Other Methods

There are several other methods available and many devices that can be purchased "off the shelf." One is a current meter, which is a propeller meter that is lowered into the stream of water and records velocity. The flow rate (cfs) is calculated by multiplying the velocity (ft/sec) by the area (ft2).

There are flumes, submerged orifices and even acoustic ultrasonic meters that use ultrasonic pulses to measure the velocity of the flow stream. All of these methods have limits to their use. For more information, refer to the Arizona Cooperative Extension publication "Measuring Water Flow and Rate on the Farm", publication AZ1130, Arizona Water Series No. 24.

Counting Tubes

If siphon tubes are used to irrigate out of an open ditch, an estimate of the flow rate can be obtained by counting the number of tubes. The size of the siphon tube and the distance from the water level in the ditch to the water level in the field (the drop) is needed to estimate the flow rate. Figure 1 shows two possible conditions. In Condition I (free flowing) the drop is the distance from the water level in the ditch to the end of the tube on the field side (usually level with the field). In Condition II (submerged), the drop is the distance from the water level in the ditch to the water level in the field. The larger the tube size or the greater the drop, the higher the flow rate.

Figure 1. Diagrams where to measure the drop distance for siphon tubes

It is often difficult to measure the difference in water levels between the ditch and the field. One easy way is to do this is to get a piece of hose and a tape measure. Put the hose in the ditch and use it to siphon water into the field. Next, slowly raise the hose in the field until the water stops coming out. Now, use your measuring tape to measure the distance between the end of the hose and the water level in the field or the outlet of an irrigation siphon tube. Make sure to keep the end up just at the level where the water stops coming out. This distance is your drop!

Measuring Water Flow in Gated Pipe

Measuring water flow in gated pipe can be accomplished many different ways. Probably the most commonly used method is the propeller meter. These meters are normally installed inside a section of pipe at the distributor's shop. The buyer then simply buys a meter section for whatever diameter pipe used. There are some other methods that can be used but for convenience and ease of measurement, the propeller is a simple and accurate method.

Propeller meters are permanent pipeline devices that measure and record the volume and flow of water moving through a pipe. The pipe must be running at full flow for the meters to operate properly. Also, there must be a straight length of pipe upstream from the meter at least 10 times the diameter of the pipe. This is to reduce the turbulence in the water it enters the meter section. Thus, a 6-inch pipe would require 60 inches of straight pipe upstream from the meter.

The meters are usually placed inside a length of aluminum pipe that is inserted into the gated pipe system. If poly-type plastic pipe is being used, there are connectors that will allow a meter section to be put in place. If you don't want to pay the expense for the meter, you can use a piece of tubing, similar to the tube method for ditches. Find a piece of tubing (preferably clear) that either fits tight inside a gate or even better, can be attached tightly to the outside of the gate. Raise the tubing into the air until the water stops flowing out. Measure the distance from the water level in the tubing to the center of the gated pipe. If clear tubing is used, then you can raise the tube well above the point when the water stops coming out and it makes for an easier measurement.

Surge Flow Surface Irrigation Technique

Surge flow is a surface irrigation technique that can save water and works well under Washington conditions. It can be defined as the intermittent application of water to furrows, rills, or borders in a series of on and off periods of constant or variable time spans. These cyclic applications, termed "hydraulic surges", are generally applied to a field through gated pipe systems or buried pipelines with furrow risers. Surge gates can be fabricated for lined concrete ditches with ports.

Increased Flexibility

Surge flow, while not the perfect surface irrigation method, does offer increased flexibility and many advantages over traditional practices. For example, growers can often complete irrigations in the same number of hours as under conventional irrigation techniques, but waste much less water. Growers can reduce variations in advance rates from furrow to furrow, and often can increase the length of run that can be irrigated with surge flow. Surge flow requires automation for economical operation.

The commercial development of controllers and valves has led to widespread adoption of surge flow in many areas in the United States with good results. It is appropriate for any land that is, or could be, surface irrigated in Washington. Surge irrigation is more easily adapted when slopes are gradual and consistent both across and down a field. An automated valve and a controller are added to a conventional application system to switch the water from one set to another at the end of a surge.

The valve controller is programmed to apply water to a given area (1 set) for a time period (usually 1 to 3 hours-based on field length and soil intake rates), then the water is diverted to a second area for an equal period of time. At the end of the second time period, the automatic valve diverts the water from the second set back to the first set. The water travels fairly quickly, with relatively little infiltration, over the already wet soil to the dry area of the furrow where most of the intake occurs during that surge. Thus, the valve alternately applies water to each of the two areas until the irrigation is completed. After water advances to the end of the field, growers can continue to surge water between sets in a cut backcycle. Since growers are pulsing water between fully advanced furrows, the rate of inflow is reduced by 50% and the corresponding tailwater also is reduced.

Advantages of Surge Flow

The advantages of surge flow surface irrigation fall into three broad categories:

- Surged water advances to the end of the field at least as rapidly as continuous flow irrigation with the same inflow rates but with a smaller volume of water, thus greatly improving the uniformity of application during the advance phase.
- Growers can reduce tailwater and deep percolation losses and can improve application efficiencies under proper automated management.
- Surge irrigation provides an inexpensive means of automating, managing, and accurately controlling the surface application of water to a field while reducing labour requirements.

The "surge" effect on surface irrigation is primarily due to reducing the soil intake rates and improving the hydraulic characteristics in the previously wetted portions of the furrows. This reduction occurs almost immediately after the initial dewatering of the soil surface. Lowering the infiltration rates may also allow a grower to slightly reduce furrow inflow rates from what was previously required and reduce soil erosion or improve water penetration on steeper lands. Light, frequent irrigations are also possible with this method, which can be advantageous when "irrigating up" a crop, irrigating shallow rooted crops, or other times when heavy water applications are not warranted.

New Idea about Surging

Surge flow is really not a new idea, as many irrigators use a similar practice by reintroducing water into a furrow or rill 12 to 48 hours after the initial irrigation to push water to the end of the run, particularly early in the spring or just after a cultivation. The new aspect is surging the water at relatively high frequencies several times a day. Surges can be alternated between 3 or more sets; however, existing commercial hardware basically limits use to two equal sets. Thus, a field is divided into an even number of sets and irrigated two sets at a time.

Uneven field shapes can be irrigated, but each set should receive equal amounts of water so that the on-farm water delivery rate remains constant. A second "delivery" pipeline is often required to supply water to a surge valve or to break a field into sets. Because different crops require different amounts of water, irrigators should design each system to cover a single crop or field rather than trying to cover several crops with the same installation. The total time for a valve to go through one on and one off cycle is referred to as the "cycle time."

The fraction of the cycle time that the water is on is called the "duty cycle." For example, a set with a 2-hour on time and a 2-hour off time has a cycle time of 4 hours and a duty cycle of 0.5. Most commercial controllers automatically increase the on times of successive surges as the water travels longer and longer distances to reach the dry sections. Selection of the proper on times for a sequence of surges will optimise advance distances of each surge down the furrow and is important in maximising efficiencies. Many controllers have preprogrammed cycle time options that will fit many field situations. The off times should probably exceed 30 minutes in most fields during water advance. The correct program for a field may change from irrigation to irrigation and from year to year. Some seasonal trial-anderror adjustments should be expected.

Increasing Efficiencies

Cutting back the furrow or rill inflow rates by 30% to 50% after water reaches the end of the field can further increase efficiencies and reduce runoff. Some valves/controllers can stop the surging and "center" the valve to split the water equally between the two sets into a cutback mode. However, doing this requires careful management. Pipelines must be able to supply water equally to all rills at the reduced flow levels.

Another option is the use of "time-averaged" cutback, where the valve is surged at cycle times as low as 10 to 20 minutes, and the averaged furrow flow rate is one-half of the instantaneous rate. The second cutback option is often preferred for pipelines lying on slopes exceeding 0.1%. Studies in Washington have shown that water can be applied more evenly by intermittent surges than by steady flow. Water under surge flow advances to the end of the field in the same elapsed clock time as continuous flow, but with half as much water. This translates into increased uniformity and water savings. Overirrigation at the top of the field is reduced.

Under proper management, application efficiencies are increased and soil erosion can be less, since the water is in the furrow for less time than in the continuous flow furrow. Also, less furrow outflow means less inflow into the tailwater and less erosion of the tailwater ditch. If surge flow is poorly managed, however, the procedure can be less efficient than use of conventional flow, and more erosion can occur. Surge flow also works better than traditional surface irrigation methods when there are crop residues in the furrows.

How to Save Water with your Irrigation System

Here are detailed instructions for things one can do that will reduce the amount of water irrigation system uses, with a few extra landscape related tricks for saving water, and a warning regarding water savings and snake oil thrown in at the end of the article.

— Have your irrigation system audited:. Some water providers will conduct an irrigation audit for you either free or at minimal cost. Check with your water provider. If not, your local water provider can likely provide a list of irrigation auditors in your area. Regardless of who does the audit, the auditor should carefully examine and test your irrigation system. They should then create a report for you detailing the condition of the system, including a list of recommendations for repairs and improvements. Some auditors also will provide you with an irrigation schedule showing how often and how long you should water during each month of the year.

— Adjust your irrigation controller (timer) run time for seasonal changes in weather once a month. Simply making a monthly change to the irrigation operation times can save more water and money than any other thing you can do. It costs nothing but a few minutes of your time each month. Most controllers even have a % key that makes changing the time quick and reasonably painless. Put a reminder on your calendar so you are reminded each month. Even greater savings come with weekly time adjustments, but monthly will provide the most return in water savings for your time invested.

— Run your irrigation system during the morning hours, especially if you use sprinklers.Less water is lost to evaporation when the temperature is cooler, plus in most areas the wind doesn't blow as hard in the mornings. Watering in the evenings can lead to turf and plant disease problems because the water sits on the plants all night, especially in humid climates.

— If you irrigate with automatic sprinklers, program your irrigation timer so that it waters in 2-3 short cycles rather than a single long period of time. Allow the water to soak in to the ground between the cycles. Almost all professional irrigation managers water their turf in cycles. For example, if you normally water for 15 minutes, try this; water for 4 minutes, wait 30 minutes or more for it to soak in, then water another 4 minutes, then wait again, then water another 4 minutes. Now you have watered a total of 12 minutes rather than 15. Even with the reduced total watering time, chances are you will see a significant improvement in how good your lawn looks. The reason cycling works so well is that almost all brands and types of sprinklers apply water much faster than it can actually soak into the ground. So after about 5 minutes of running, most of the water begins to build up on top of the soil and then it just runs off into the gutter or to a low spot in the yard. Cycling the irrigation gives the water time to soak into the ground and reduces water run-off, it also will help reduce the wet spots in the lawn where lawn diseases get started.

— Make sure tall grass, groundcovers, or shrubs are not blocking or deflecting the water spraying out of the sprinklers. The water from sprinklers heads that pop-up less than 3 inches high is often deflected by tall grass around the sprinkler head. When the water pattern is deflected by tall grass or leaves it results in uneven watering and water waste. With tall grass it may appear that the water spray is forcing it's way through the grass without being deflected, but this is an illusion. You can only see the large, heavier drops of water that are not easily deflected. The smaller droplets that you can't see are being blocked, and that creates uneven watering patterns. The industry standard for lawns is to use sprinkler heads with a pop-up height of 4 inches or more. Turn on the sprinklers right before the next time the lawn is scheduled to be mowed and see if the grass around the heads is blocking the spray. If it is, consider replacing the sprinkler heads with a model that pops up higher.

— Shrubs and groundcover that have grown since the sprinkler system was installed may also block the spray of sprinklers. If you don't want to replace or raise the sprinkler heads, trim the shrubs around the heads so that the spray is not blocked. In shrub areas it is not always necessary for the spray to go over the top of the shrubs. In many cases it is OK for the water to spray into the side of the shrubs, especially if the shrubs are 6 feet (2m) or more away from the sprinkler. Shrub roots will often grow out to where the water is. If the shrubs are not wilting and are healthy, then there is no need to change the sprinklers. If you want to really save a lot of water consider changing the sprinklers in shrub areas to a drip system, which will use even less water. More information on drip systems is found farther down in this article.

— Relocate sprinklers so that they are between 4 and 6 inches (10-15cm) from the edge of sidewalks, curbs, patios, etc. in lawn areas. In shrub areas they can often be 12 inches (30cm) from the edge, especially with a mature landscape. This will reduce the amount of spray onto the paved surface and will not create a dry area along the edge of the lawn. It will also reduce the amount of damage that trimmers cause to the sprinkler heads. Almost all stores that sell irrigation equipment will have flexible riser pipes made for relocating sprinklers. Using the flexible riser pipe makes relocating the sprinklers much easier, and the flexible pipe allows the sprinklers to move if a car or heavy lawn mower hits them without breaking a pipe or the sprinkler.

— Fix leaking valves. Look for water running onto sidewalks or over curbs after the sprinkler system is turned off. If water flows constantly when the sprinkler system is off (often there will be mold or algae growing on the cement or ground) that indicates that a valve is not fully closing. A valve that doesn't close usually is caused by a small grain of sand stuck inside the valve. Clean, or simply replace, the valve.

— Fix low head drainage. Do your sprinklers spit and spew air mixed with water for a short period each time they are turned on? This is caused by a phenomena called "low head drainage". Low-head drainage occurs when the sprinkler system has been installed on a sloped area. After the sprinklers are turned off, the water in the pipes drains out through the lowest sprinkler heads and is replaced by air. The water that drains out is wasted, and often flows into the gutter or creates a muddy area around the lowest sprinkler head or drip emitter. Then the air is violently forced out the next time you run the sprinklers. This puts a lot of stress on the sprinklers and pipes. The easiest way to tell if you have this problem is when you turn on the sprinklers. If they spit and spew air when the valve is turned on, then you have low head drainage.

— Install a Smart controller. A Smart controller does the work of periodically adjusting the sprinkler operating times for you. It changes the run times to reflect the current water needs of the plants. Some water companies will assist in the purchase price of a Smart controller..

— Install a rain switch. A rain switch is a simple rain sensor. When it detects measurable rainfall, it turns off the automatic irrigation valves. You can buy a rain switch almost anywhere irrigation products are sold, most will work with any brand of irrigation controller or timer and any brand of valve. You mount the rain switch on the side of the house, on a pole, or on a fence in a location where water will fall on it but sprinkler water will not hit it. Then you run 2 or 3 wires from the rain switch to the controller. The exact installation instructions vary depending on the brand and model of the rain switch.

— Install a filter on your irrigation system. A filter saves water (and money) indirectly. Most valve and sprinkler malfunctions result from contaminants in the water supply. Typically this is small grains of sand, pipe scale, or small fresh-water snails. All of these are common in many public water systems. Installing a simple screen filter at the water source (before the valves) will greatly reduce the frequency of sprinkler system breakdowns and save water. A filter is one addition to your irrigation system that almost always pays for itself within 5 years. The cost of a single valve repair can be much greater than the cost of buying and installing a filter.

— If your irrigation system is located in an area where hard frosts occur make sure you properly winterize it each year before the cold weather hits.

— Switch to newer sprinkler heads. Technology in sprinklers has advanced over the last 20 years and many new sprinklers are more water efficient than the older models. Generally this option is only cost effective if you have a very old sprinkler system, or if your original sprinkler system was poorly designed. Small stream-rotor nozzles that fit onto spray-type sprinkler bodies are being heavily advertised as being more efficient than old style spray heads. A couple of popular brands of these stream rotor nozzles are the Hunter MP Rotator and the Rainbird Rotary

Nozzles. A word of warning; while they are indeed more efficient than a standard spray nozzle, it is only a marginal efficiency gain and switching nozzles is probably not cost effective if you have a good sprinkler system. However these rotary nozzles can provide a significant benefit for poorly designed sprinkler systems where spray type heads were spaced too far apart or the pipes are too small. This is because these new rotary nozzles achieve a greater radius than was possible with the old style nozzles while using less water. So if you have a sprinkler system and you give it a tune-up, but it still has dry spots, changing to the stream rotor nozzles may help. Measure how far apart your sprinkler heads are, then select the rotary nozzle that has a radius equal too the distance between sprinklers. So if the sprinklers are 20 feet apart get nozzles that have a 20 foot radius. Replace all of the nozzles, don't try to mix the rotary nozzles on the same system with older nozzles- they are not compatible. It may help, and it may not. You might want to buy the nozzles at a the store with a generous return policy. That way if they don't fix the problem you can return them. Not all spacing problems can be repaired using stream rotor nozzles. You may need to add more sprinklers, or just totally redesign the sprinkler system to fix the dry spots.

— Switch to drip irrigation for watering shrubs. Drip irrigation is about 20% more water efficient than sprinklers are. It is easy to install, and reasonably inexpensive.

— Do you have an alternate source of irrigation water you could use? Water from creeks, ponds, and shallow wells are all examples. Grey water from roofs and sinks is another source if you have very limited irrigation needs. These are typically not easy solutions. You need to do a lot of research before you do anything. Completely design out the system and make sure you have enough water- and make sure you have a legal right to the water! There are way too many tanks, barrels, and pumps sitting around unused because someone got all excited and bought stuff without researching how much water was really required. Also, digging your own well is not legal in many locations, and requires a government permit to do it pretty much everywhere. If you punch an illegal well out in your yard and it pollutes the aquifer you could lose everything you own paying for the environmental damage. If you dam up a creek and kill some endangered whatever you could find yourself in a lot of trouble. There was a time when you could do whatever you want on your own property, but those days are long gone. Right or not, that is the way it is.

— Some sprinkler head models have built-in pressure regulators. The pressure regulators save water by reducing the water pressure at the sprinkler head nozzle. If too much water pressure is present, the sprinklers tend to create too much mist and give uneven coverage resulting in water waste. These built-in pressure regulators are available as an option on many higher quality spray-type sprinklers. Rotor-type sprinklers seldom need pressure regulation, so this feature is not generally available on them. In order for the pressure regulators to work you must have excessive pressure! If you do not have excess pressure, the pressure regulators may actually harm your system's performance. If you are a homeowner and considering retrofitting your sprinkler system using pressure regulating sprinkler heads, turn on the sprinklers and look carefully at the sprinkler head farthest from the valve. If it is not creating a considerable amount of mist, it is unlikely that the pressure regulators will help you save water. So what would be a typical situation where pressure regulating sprinklers would be used? A typical use of pressure regulating sprinkler heads is for a valve circuit on a steep hillside. In this situation the sprinklers at the bottom of the hill will have excessive pressure due to the effect of gravity on the water pressure (take my word for it, this is too complex a topic to cover here.) Using pressure regulating sprinklers on these lower sprinkler heads would cause them to perform better. If the highest sprinkler is not more than 6 feet (2m) higher than the lowest sprinkler on the same valve circuit, pressure regulating sprinklers will probably not be of much help.

— On a typical sprinkler system (that is not on a hillside) you can get almost the same result obtained from switching to pressure regulating sprinkler heads by simply properly adjusting the system. Try properly adjusting the sprinklers first, before you spend a lot of money on pressure regulating sprinkler heads. My experience is that very few sprinkler systems will benefit significantly from the use of the pressure reducing sprinkler heads. If you do have too much water pressure it is usually better to install a single pressure reducer valve on the entire irrigation system, or install pressure reducing valves. Excessive pressure is damaging to the entire sprinkler system, so it is better, and often cheaper, to reduce the pressure in the whole system.

— Automated emergency shut-off devices save water by automatically shutting off the water when something in the irrigation system breaks. There are several different types of these devices available. Before we get into specifics, understand that these devices do not save water during ordinary operation of the irrigation system. They only prevent water waste when something breaks. In many cases they simply are not cost effective, so the suitability of these devices must be examined on a case-by-case basis. Automated emergency shut-off devices are often used on irrigation systems where a break or valve failure could cause serious damage They also are used often in locations where a leak might go undetected for days, such as a vacation home or remote location.

The first of these devices is a irrigation controller with the capability of monitoring flows. The controller works in conjunction with a flow sensor and master valve. The master valve is an electric solenoid valve that is installed on the mainline near the water connection point, where it can shut-off all the water flowing to the entire irrigation system. The master valve is typically installed right after the backflow preventer. The flow sensor is installed on the mainline pipe after the master valve. Be sure to follow the manufacturer's directions when installing a flow sensor as it will not work accurately if it is not installed exactly as recommended. The flow sensor is connected by wires to the irrigation controller. Some controllers use a wireless signal to communicate with the flow sensor. The controller used must be a model that is capable of monitoring the flow sensor and responding to the input from it. Typically this feature is only found on higher-end controllers. If you're interested in this type of system, select the controller first. The controller manufacturer will then have specific recommendations as to the type and model of flow sensor and master valve to use.

The second type of shut-off devices are simple mechanical devices that either install under, or are built into, a sprinkler head. These are often called "geyser preventers", a reference to the geyser-like spray of water that occurs when a sprinkler head or nozzle is broken off. These devices are flow activated. If the sprinkler head or nozzle should break off, it will result in a much higher water flow. The higher velocity of this increased flow through the shut-off device causes a valve in the device to close, which shuts off the water supply to the broken sprinkler. Some shut off devices screw directly into the pipe tee under the sprinkler head. The sprinkler riser is then screwed into the shut-off device, and the sprinkler is attached to the riser. Another type of shut-off devices are built into the bottom of some models of sprinkler heads and are often called "valve-in-stem shut-off devices." The limitations of the built-in devices is that since they are inside the sprinkler they will not work if the entire sprinkler head breaks off, which is a common problem. Keep in mind that neither of these types of mechanical shut-off devices will detect a broken pipe, they only work if the sprinkler breaks. Also be sure to investigate carefully the details on any shut-off device you're considering purchasing. While some devices completely shut off the flow when a break is detected, others simply restrict the flow, but do not completely shut the water off when the sprinkler breaks.

— Separate plants into hydro-zones. A hydro-zone is an area where all the plants use more or less the same amount of water and have the same sun and wind exposure. For example, lawn in the sun would be one hydro-zone, the lawn in shaded areas would be another hydro-zone, lawn in the sun on a windy hill-top would be yet another hydro-zone. The irrigation is separated so that each hydro-zone area is watered by a different valve. This allows you to water each hydro-zone individually for just the right time to apply the water needed by the plants, without over-watering.

Landscape Ideas for Saving Water:

— Mow your grass at a higher length (so that it is longer.) While there is some debate about whether this saves much water, scalping the grass off at a low height is definitely not good for the vigor and health of the grass. Longer grass has deeper, stronger roots and is more resistant to disease and drought. Most grass should be mowed to a length of no less than 3 inches.

— Dethatch and/or aerate your lawn. Lawn aeration helps assure that the water can penetrate easily into the soil, over time the soil surface can become very compacted and water will not easily penetrate it. Aerating also provides air to the roots of the grass, which is necessary for healthy growth. Thatch build-up on the soil surface under the grass blades can actually repel water. How often you need to dethatch or aerate the lawn depends on the type of grass, whether you remove lawn clippings, the type of soil, the climate, and how much you fertilize. Dry spots at the higher areas of the lawn are often the first sign you need to dethatch or aerate the lawn. To check take a hose and lightly water the dry area. If the water does not penetrate into the soil quickly you need to either aerate or dethatch, or both.

— Reduce the use of fertilizers. Fertilizers encourage rapid growth which results in higher water use. Cut back on fertilizer application amounts to the minimum needed. More frequent application of fertilizer in smaller doses will also help. Try to avoid the green up then yellow off then green up cycle of fertilizer application. Some people find that using an automatic fertilizer dispensing system gives them the green yard they want without wasting fertilizer. Fertilizer injection isn't for everyone, you need to be willing to calibrate the system and keep watch on it.

— Add a layer of mulch to shrub beds. A 2 or 3 inches deep layer of mulch, such as wood chips, bark, almond hulls, or even decorative rock, reduces water use and also reduces the number of weeds.

— Reshape your landscape to use less water. Often a minor change can not only refresh and improve the appearance of your landscape, it can also save water. Look around at your yard layout, especially the size and location of lawns. Can you remove or shrink the size of the lawn areas? Lawn uses much more water than the same size area planted in shrubs or groundcover. Does your lawn go right up to the edge of the house or fences? If it does, you can save water and help your house siding and fences to last longer by reducing the size of the lawn so that it is at least 3 to 4 feet away from fences and walls. Irrigation water spraying on the side of a house or fence can cause all kinds of expensive problems. Try creating a curved meandering edge on the lawn area to mimic the look of a meadow; it gives a much more esthetically pleasing look than a straight edge. Then add a foundation planting of low-growing shrubs with drip irrigation between the lawn and the house or fence. For more design ideas go down to the local hardware or book store and pick up a couple of books on landscape design.

— Have you looked at the new synthetic lawns and golf greens? Many of them look very good and are very durable, they are much improved from the old fake grass "carpets" of the past. Synthetic grass isn't going to meet the needs of everyone, but they sure save a lot of water compared to a real grass lawn!

— How about replacing old high-water using shrubs with shrubs that are less thirsty? Visit a local nursery with knowledgeable staff who can help you select good plants that use less water for your yard. Use of native plants can be particularly water and environmentally friendly, but isn't necessary to get a water saving landscape. You also don't need to go to a "weeds and twigs" look just to save water. While a desert landscape may use almost no water, even a lush-looking landscape that mimics a forest can be designed to use minimal water. If you want a small garden area of high water use plants, locate it in a shady area with protection from wind. Even high-water-use plants will use much less water if they are planted in a shady area where they are protected from strong winds.

Water Savings & Snake Oil

As attention has shifted to saving water, irrigation companies have assigned their marketing departments and sales staff to push for any water saving connection possible that will sell products. "Irrigation specialists" are suddenly everywhere offering services to help you revamp your irrigation system to save water. Unfortunately, along with the many legitimate products and companies a few "snake oil" salesmen are bound to sneak in, trying to turn a quick profit at your expense. In the past few months I have seen a number of questionable claims. Before you fork over your money or sign on the line, ask a few questions:

— How does this product work? What feature about it saves water and how? An ad I recently saw in an irrigation trade magazine promoted a product as "water saving" but failed to disclose that the product would save water only in very specific situations found on very few irrigation systems.

— Will this product work with your irrigation system? Will it fit? A client of mine was recently told (by a city agency) he should switch his spray-type sprinklers to the new stream rotor nozzles to save water. These stream rotor nozzles have a minimum radius of 10 feet. His sprinkler system uses sprinklers with a radius of 8 feet and less. If he had made the requested change it would not have saved any water, it would have resulted in massive water waste!

— Does the firm proposing this service have extensive irrigation experience and knowledge? A lot of landscapers who have never installed a sprinkler are suddenly irrigation experts.

— Be wary of claims that you will save some large percentage of water. Most of the claims I have seen of 50-80% savings were based on the assumption that you have a really terrible quality irrigation system and that you leave it set to water for the maximum amount all year, even when snow is on the ground. In that case simply turning it off in winter could net you a 50% savings. So beware of blanket claims.

References

Burt, C.M., Robb, G.A. and Hanon, A., *Rapid evaluation of furrow irrigation efficiencies*, Paper 82-2537 presented at the Winter Meeting of ASAE, Chicago, Illinois, 1982.

FAO., "Surface irrigation," by L.J. Booher, *FAO Agricultural Development Paper No. 95*, Rome, 1974,160p.

Humpherys, A.S., "Mechanical structures for farm irrigation," *J. Irrig. and Drainage Div.*, ASCE, 95(IR4):463-479, 1969.

James, L.G., J.M. Erpenbeck, D.L. Bassett and J.E. Middleton., *Irrigation Requirements for Washington: Estimates and Methodology,* Washington State University Cooperative Extension Bulletin EB 1513, 1989.

Ley, T.W., *Simple Irrigation Scheduling Using Pan Evaporation,* Washington State University Cooperative Extension Bulletin EB1304, 1995.

Trimmer W., and H. Hansen., *Irrigation Scheduling,* Pacific Northwest Extension Publication 288, 1994.

6

Water Conservation in Semi-arid Areas

Agricultural development naturally takes place first on the best land. Whether at the scale of the individual farm or a whole country, the tendency is to use the best land first. When there is a need to increase agricultural production it is usually directed to maximizing production in the areas which have the best potential. But as demand increases for the products of the land - food, fuel, shelter, and clothing - it is necessary to make increasing use of land which is less suitable for agriculture, or land in less favourable climates. People concerned with agricultural planning, or development, or production, must all pay more attention to the semi-arid regions.

Attention has been sharpened by the widespread droughts in the early nineteen-seventies, and again in the mid-eighties, as shown by the surge of conferences and workshops in the last ten years, and the explosion of aid programmes in semi-arid areas.

It is futile to expect magic solutions to the problems of semi-arid regions. We cannot alter the unreliability of the rainfall, nor the fact that unpredictable rainfall and the occurrence of droughts are inevitable. Neither can centuries of abuse and mismanagement of the land and the people be corrected overnight. Short-term results to reduce starvation through improved yields is only part of the story; a programme to win the confidence of subsistence farmers should be planned to a timespan of at least five years, and plans to direct the attitudes of governments more towards land use will need much longer.

Drought and Change

The place of droughts in relation to food supplies was admirably expressed by Sir Joseph Hutchinson in the preface to the Royal Society Symposium of 1977, when he said:

> "These problems (widespread droughts of the early nineteen seventies) were not created by drought. Drought is but the dry extreme of the natural climate variability of semi-arid areas. The problems arise from the pressure of human and domestic animal populations on limited resources, and they have been aggravated by the rapid explosion of populations in recent years....
>
> "It is in the nature of these marginal areas that agriculturally disastrous seasons are within the normal range of climatic variations. Such seasons must be accepted as something to be foreseen and planned for, and not as an "act of God" to be met by international charities. The difficulty of planning for drought conditions and hence of preventing famine has been increased by the great successes of science and technology in the last quarter century in the control of parasitic and nutritional disease.
>
> Improvement in health has led to great increases in human and animal populations, and these have been maintained by fuller exploitation of the environment. Thus the pressure at the margin has increased, and what was a shortage that could be met by encroachment on unused resources becomes a famine because no unused resources remain."

The all too common attitude towards drought is typified by the Australian outback farmer who is reported to have explained his problems to the extension officer because "we have not had a normal rainy season for 25 years".

Hutchinson's comment about increasing pressure at the margins is reinforced by Ormerod who is concerned about the problems which may inadvertently arise from economic development in dry areas, and argues that "One of the most important factors operating to increase aridity in West Africa is the economic demand which has stimulated the growth of herds and of arable farming which compete for land in the arid range areas."

Ormerod suggests that the conventional view that demand is best satisfied by increasing local production may increase the rate of agricultural degradation and goes on to say:

> "There is a particular danger in stimulating economic advance in desert areas and particularly in linking nomads to the monetary system and I make the plea that, before barriers to development... in this remote and fragile part

> of the world are broken down, more information should be obtained about the dynamics of nomadic grazing and of the possible effects that the extension of their operations might have upon the environment and the climate........"

He also urges the development of quantitative techniques which could be used to assess the capacity of soils to withstand exploitation, to provide a measure of the limits to which economic development can be pressed without causing ecological degradation.

The Extent of Erosion

Most semi-arid regions suffer severe rainfall erosion, often more than in the humid tropics, because the rain has a high erosive capacity which is more damaging because, of the reduced vegetative protection Past erosion in Mediterranean lands is well documented by Lowdermilk and a recent survey shows that the position continues to deteriorate. Some examples are Morocco, with 40 percent of the country's total land area exposed to erosion; Turkey with more than 50 percent damaged; Greece with 21 percent suffering moderate erosion and 46 percent severe erosion damage.

The non-Mediterranean semi-arid regions also report rainfall erosion problems; in North America, the USA and Mexico; in South America, the north-east corner of Brazil, and much of the south-east of the continent; in Africa, the Sahel and the south-west (e.g. Botswana); then there is a huge belt stretching east from Africa through Turkey, Iraq, Iran, and Afghanistan, with one arm extending into eastern USSR and China, and the other arm south of the Himalaya through Pakistan to India. In Australia, most of the states have severe rainfall erosion problems in areas of low rainfall.

Wind erosion can be serious in the temperate semi-arid regions and examples are the central steppes of Asia (eastern USSR, Mongolia, and China); in North America the north-western States of the USA, and the central prairies of Canada. In the semi-arid tropics the picture varies. Some countries surrounding the arid Sahara report desertification with increasing wind erosion. On the other hand, the opinion has been expressed that in many cases the semi-arid regions receive more soil blown by the wind from deserts than they lose from rainfall erosion. This view is supported by the existence of vast areas of wind-deposited loess soil in semi-arid regions.

Differences Between Semi-arid Regions

The only thing common to all the semi-arid regions is low rainfall. Everything else varies so much that it is not practical to think in terms of searching for universal solutions. They may have temperate or tropical climates, rainfall in summer or winter seasons, and hence totally different forms of agriculture.

Even when the climate is similar there are other significant differences as shown when Charreau considered the possibilities of transferring the technology developed at ICRISAT in India to countries in West Africa. The main features of the climate are the same in both regions: a short rainy season, intensive rainfall interspersed with unpredictable droughts, and highly variable rainfall during the wet season. However, in India the duration of the rainy season is longer, and the curve of potential evapotranspiration is unimodal, while it is bimodal in West Africa. Non-typical periods without rain are most likely to occur in the middle of the rainy season in India, but at the beginning and the end of the season in West Africa. Both regions have, as their main soils, red alfisols and black vertisols, but there are differences which have important implications for cropping, cultivation, drainage, and fertility status. In addition there are major sociological differences, particularly the density of population which in India is very high at 200 inhabitants per km^2 compared to 10-60 per km^2 in semi-arid West Africa. This leads to a very different approach to agricultural production. In West Africa the objective is to maximize production per worker, whereas in India it is more important to maximize the return per unit of land. There are also differences in the style of land use. In Asia families have farmed the same land for generations; in Africa many people are nomadic or seasonally migrant.

Shortage of Information

Previously the emphasis of agricultural research and development has been on making the best use of good soils and good climates and little attention was given to marginal environments. One result of this uneven spread of effort is a lack of information about the less favoured semi-arid areas. This shows up in every technical discipline. In the semi-arid regions there are fewer soil surveys to show what is available, and less research on the physical and chemical properties of soils to show their capablities and problems. There are fewer agrometeorological stations, and less analyses of the results, although because of the variability there really should be more of both.

There has been less research on crops. Most countries with a combination of humid and arid conditions have concentrated on the humid areas, sensibly because this is likely to give the quickest and most cost-effective return. Major international research for dry areas is recent, with the establishment of the International Crops Research Institute for the Semi-Arid Tropics (ICRISAT) in 1972 and the International Centre For Agricultural Research in the Dry Areas (ICARDA) in 1977. Selecting or breeding cultivars for particular semi-arid climates shows promise, but has a long way to go.

There is also a lack of information on the social and economic side. The farming systems, and the economics of agricultural production, have been studied less, partly as a result of the lower interest, partly because of the physical difficulties of large areas and fewer roads. Also the farming systems are complex because of:

— combinations of settled agriculture and nomadic or semi-nomadic livestock, or seasonal transhumance;
— cultural complexities of communal ownership or tribal grazing rights;
— cultural and ethnic boundaries may be different from the political boundaries.

Lack of Technology

A combination of lack of interest, low research commitment, and the complexities of the problem results in a shortage of technology which can be applied to improving agriculture in semi-arid regions. Jones points out that, in Africa, studies of water management techniques such as tillage and tied-ridging have several times shown great promise, but have been too brief, or too local, or restricted to experiment stations, and as a result they "generate neither the depth and breadth of understanding of the technique (and the limitations imposed on it by soil, implements, economics, etc.) nor the impetus and general interest that might sustain it through a long programme of wider testing, adjustment, and retesting under practical farming conditions". Research in semi-arid regions needs a long time span to reduce the effect of seasonal variations, and coverage over a wide area to reduce the effect of local variations.

Another difficulty is putting technology into widespread operation after it has been developed and tested. In India, agricultural production is unstable in over three-quarters of the agricultural areas because of inadequate or erratic rainfall. The Drought-prone Areas Programme has made considerable progress over twenty years, but there is still a gap between the research results and practical application to farming. Improved technology for the vertisols, developed at ICRISAT, has been tested and validated, and is being increasingly taken up. But to get it applied on the whole 5 million hectares of vertisols within 20 years would require a growth rate of 43 percent per year in the rate of adoption.

An important reason for the poor adoption of new techniques is the inability of the subsistence farmer to take risks. The essence of farming is trying to improve the odds in the gamble against weather, pest, and disease. The peasant has no risk capital to gamble with, so his whole strategy is geared to safety. He would rather use a low-yielding variety which gives some yield every year than an improved variety which will give an increased yield most years, but none at all in the bad year. Even if the chance of an increased yield is nine years out of ten, it is still not an acceptable gamble for the small subsistence farmer. In the tenth year, the year of failure, his family will starve. It is neither stupidity nor lethargy when he sticks to his old variety, it is accepting the realities of life.

Pressures

The human population of semi-arid regions is increasing like any other region as a result of natural increase, but there are other factors. There is a general migration from more densely settled areas, in some cases encouraged by national policies of land allocation or for improving production in the drier regions. It has also been suggested that the increase of settled agriculture in areas previously occupied by nomadic pastoralists has been accompanied by a higher growth rate among the settled population than the nomads. Increases in the livestock population are sufficiently well documented in many countries for us to be able to recognize a trend of steady increase, usually an upward exponential curve.

An interesting example is the sheep population in Iraq as quoted by Thalen, who suggests that a remarkable increase in the sheep population resulted from a combination of the introduction of mechanical equipment for drilling wells and the adoption of four-wheel-drive vehicles which greatly increased the accessibility of remote regions. The national herd was estimated at 4 to 5 million in the nineteen-thirties and rose to 10 or 11 million in the nineteen-sixties, leading to such overgrazing and degradation that the population fell back to 6 million in the early nineteen-seventies. Similar examples of over usage following improved water supplies are reported from the Sudan and from other countries in the Sahel.

Increased pressure on the natural resources also leads to loss of flexibility. There are no reserves which can be called upon in the bad year. This applies at the scale of the individual farmer whose meagre reserves of food and grazing are soon used; it applies equally at the level of the village or tribe who in the past have traditionally held emergency grazing in reserve; it applies at the national level as grain stocks are depleted, and at the regional level when it is no longer possible to use the safety valve of major movement of livestock. Examples of traditional safety valves are Botswana, where a poor season is the occasion for the large movement of stock out to the east of the country, and the Sahel where there is a general movement southward, in both cases with a reverse migration as conditions improve. In western Sudan the effect of recent droughts has not been so severe because the people were able to move south with their livestock to wetter areas which are only sparsely populated.

Political events may provide restraints on traditional movements of this nature by restricting the movement across national frontiers. On the other hand, political troubles can also lead to problems caused by the movement of refugees.

The general trend of change in land use has been quantified in the case of Mali. Between 1952 and 1975 savanna grazing went down from 89 to 63 percent of the farmed land while cropping rose from 6.6 to 11.1 percent. At the same time degraded land increased from 4 to 26 percent.

To summarize, there is a wide range of evidence to show that the natural resources of the semi-arid regions are suffering increasing damage as they face increasing demands and increasing pressures.

Climate

Rainfall

The two key features which adversely affect agriculture are the low amounts of rainfall and the unreliability. Adding to the problem, the two are linked, in that the variability becomes greater as the mean rainfall decreases. Cropping may not be possible at all, or possible only with the use of special techniques such as fallows, or furrows to collect more rain. The range of suitable species will be limited, and in the semi-arid tropics with summer rainfall it is usual to grow maize where the rainfall permits, then sorghum where the rain is not sufficient for maize, and millet where the rain is not sufficient for sorghum. Farming strategy must therefore be directed towards minimizing any loss or wastage of rainfall.

There is a general association between mean annual rainfall and the annual erosive power, but this results mainly from the fact that a higher annual rainfall includes a greater number of storms, not that the individual storms are more severe. The high-intensity, short-duration convective rainfall dominant in the tropical semi-arid regions has fairly stable intensity/duration relationships, and these are independent of the longterm rainfall at a particular location. For example, an important parameter for the prediction of erosion is the maximum fall in a 30-minute period and this has been shown to be independent of location and of mean annual rainfall both in Mediterranean lands and the semi-arid tropics. On the other hand, some areas experience low-intensity frontal-type rains, usually in the winter season. Where most precipitation occurs in winter, as in Jordan and in the Negev, low-intensity rainfall may represent the greater part of annual rainfall, and rainfall erosion is generally less in these circumstances, although sensational floods can still occur.

In temperate climates, in nineteen years out of twenty, annual rainfall is between 75 and 125 percent of the mean. In the semi-arid tropics, at mean annual rainfalls of 200-300 mm, the rainfall in nineteen years out of twenty typically ranges from 40 to 200 percent of the mean, and for annual rainfalls of 100 mm the range widens and is 30 to 350 percent of the mean.

There is usually a considerable positive skewness of the distribution of annual point rainfalls. This means that there are more annual values below the mean than above it, and the modal value is less than the median value, and both the mode and the median are less than the mean. There is also the question of short-term spatial variation which is common when the rainfall is mainly of the convective small-cell type. For example, Sharon showed that in Israel the rain may be in small cells of up to 5 km diameter, so that in a watershed of a few hundred km^2 only 20 percent of the area may receive rain.

In other words, the general unreliability of rainfall from year to year is compounded by the variations from place to place and from storm to storm. Forecasting or prediction is therefore also difficult. Modern statistical techniques allow the calculation of the probability of events important to agriculture such as the occurrence of planting rains, or of drought spells at particular times during the season. But these are essentially strategic tools, helpful to governments for planning policies, or offering guidance on the probability of being able to carry out crop management practices at particular times. There are promising possibilities of providing seasonal forecasts for a region, based on links between large-scale features of the atmospheric circulation and rainfall, for example a study of Fiji by Dennett et al., but such forecasts can only be general and will show considerable local variability. What the farmer really needs is advice to help him in short-term tactical decisions, but this is more difficult. Studies by Dennett et al. found no dependence between rainfall at different times within the rainy season; that is, events early in the season do not provide any guide on what is likely to happen later in the season. On the other hand, in Kenya, Stewart and Hash were able to assign a rainy season into one of three categories of predicted total rainfall, according to the time and early amount of the early rains, and to do this sufficiently early to modify management decisions. But a similar analysis of Botswana rainfall data revealed no linkage between early rainfall and seasonal total.

There is thus some opportunity for tactical mid-season changes. In another example, Rainey observes that nomadic graziers respond quickly and effectively to seasonal variations in rainfall by moving herds to where their bush telegraph intelligence system tells them that rain has fallen. Mid-season changes of cropping patterns are more difficult. Theoretically it should be possible to plant a preferred long duration variety if the rains start on time, but switch to a short duration crop if the rains are late. There are some cases of this technique being used, for example in India and in Swaziland, but there are serious logistic problems of maintaining duplicate seed stocks and of distributing them at the required time.

Temperature and Wind

Extremes of temperature can cause problems at both ends of the scale. The wheat growing areas of Kazakhstan, USSR, are only free from frost for 100-120 days, and cropping is only made possible by day and night ploughing, planting as early as possible, and harvesting in autumn before the crop is fully mature Botswana in Southern Africa experiences surface temperatures so high that germination can be inhibited, and it is possible that one reason for reduced yields after ridging is an increased soil temperature in the ridges.

Wind erosion is likely to be a threat whenever soil dryness and wind occur together. This combination is frequent in the semi-arid tropics, and can also occur in temperate climates when seasonal dryness coincides with seasonal winds.

Hot dry winds have three main effects. They reduce the effectiveness of rainfall by evaporation from the soil surface; they increase evapotranspiration from the leaf area of crops, increasing the risk of moisture stress; and the surface of stored water suffers from high evaporation loss.

Soil

The diversity of soils in semi-arid regions is immense, and two examples will serve to indicate the range of soils and soil-related problems. In a review of the soils of the semi-arid tropics, Kampen and Burford show that the semi-arid tropics are mainly in the developing countries of Africa (70 percent of the total) and south-east Asia (mostly in India). Latin America and Australia each contain about 10 percent. Eiqht soil orders are represented, of which alfisols and aridisols are more than half of the total. Alfisols are the largest order and cover about 32 percent of the African SAT and about 38 percent of the Asian SAT. Vertisols cover 6 percent of the total area but 25 percent of semi-arid India.

Another example is Jones, explaining the diversity of soils in the low rainfall areas of southern Africa, "the soils vary widely from very extensive areas of arenosols (sands and loamy sands) in Botswana, Zimbabwe and Mozambique, to smaller but potentially important areas of vertisols (heavy clays) scattered widely over the zone; and from highly leached and acid acrisols, ferralsols, and nitosols in Tanzania (mainly) to neutral, alkaline, and salt-affected soils in Botswana (fluvisols, solonchaks, and xerosols and solonetz)."

Physical properties which adversely affect agriculture on these soils are equally variable. Some examples are:

— low rates of infiltration leading to high run-off and hence less effective utilization of the rainfall;

— surface crusting is widespread in semi-arid regions and may be the primary reason for low infiltration. Apart from the reduction in infiltration, surface crusting may hinder the emergence of seedlings to the point where it has to be included as a factor in capability classification, as in Zimbabwe;

— dense packing of soils may reduce infiltration and also increase the draught requirement for tillage operations, as in the 'hardveld' sandy loams in Botswana;

— the deep cracking of vertisols can lead to increased loss of moisture by evaporation, and to cultivation problems;

— many semi-arid soils have low moisture-holding capacity, especially shallow or sandy soils. The shallow alfisols in India are an example.

Water storage in the root zone may be limited by a) low intrinsic moisture-holding capacity of sandy soils and b) the limitation on root proliferation at depth by natural soil hardness. Management techniques to prevent run-off may cause ponding and over-wet conditions particularly at early stages of growth. Due to erratic rainfall and low storage, even complete infiltration early in the season may not avoid moisture stress later.

Chemical soil problems include low fertility which may be inherent or caused by leaching or by past soil erosion. Breman and Vithol report that, in the Sahelian rangelands, the low fertility of soils, especially in nitrogen and phosphate, is often more of a limiting factor than low and irregular rainfall.

There are a number of specialized forms of erosion which are associated with excessively high erodibility and which are more frequent in semi-arid than in humid regions. Tunnel erosion or piping is one of these and often associated with badlands geomorphology. This may be associated with dispersing soils or soils with a calcium/sodium imbalance.

The Farming Background

All the problems of developing countries are present in the semiarid regions, but usually intensified by the low levels of production, which lead to little or no investment capacity on the farm, or low development of the infrastructure by government. The general picture is rather depressing with declining production and increasing degradation, but examples show that techniques exist to overcome the problem.

i. In the semi-arid Pacific north-west of the USA, successful wheat production has been achieved on mean annual rainfall of 240 400 mm but this is mainly winter rainfall, and the system requires substantial inputs of machinery, capital, and fertilizer.

ii. Turkey is another success story, where the annual yield of wheat rose from 8 million tonnes in 1965 to 16 million tonnes in 1976 as the result of a programme which included:
   - research programmes on tillage, weed control, and breeding;
   - a substantial increase in the national extension and training programme;
   - advanced training overseas for research and extension workers. In this case also there were substantial inputs but they consisted of training, research and extension, so the example is more relevant and applicable to developing countries.

iii. In Australia, reasonable levels of yield are being maintained through the use of clover or medic rotations. The only fertilizer used is superphosphate at 60 to 100 kg/ha. Grazing of the legumes and stover has to be carefully controlled. This system has been successfully demonstrated in Libya, and has potential for other areas.

iv. In summer rainfall areas there are promising developments of dry-land farming techniques coming from ICRISAT, although farmer acceptance is still limited.

In many countries a serious constraint is the shortage of energy available for cultivation. In the third world, the poorest farmers can only dream of the day when they may be able to own a pair of oxen, and millions more have some animals but not enough to work their land effectively. In summer rainfall areas there is always the problem that in spring, when there is the greatest need for draught power, and ploughing has the highest draught requirement, the animals are in poor condition after coming through a hard dry winter. It is common in Africa to find that after the rains start, the animals have to be fed up on the new grass for a month to get them into condition, although early planting is one of the most basic requirements for good crop production. In other countries the critical shortage may be manpower. Migration of adult male labour to more rewarding occupations in towns or in neighbouring countries has a serious effect on agriculture in many countries. In Africa, the gold mines and industry of South Africa have drawn away so much labour from neighbouring Lesotho and Swaziland that crop production is materially reduced. Another recent market for unskilled labour has sprung up in the oil-rich Middle East. Migration from densely populated countries like India and Pakistan does not affect their agriculture, but in other cases there are problems. In the Yemen Arab Republic a well-developed system of terracing is collapsing because of inadequate labour to maintain it. A similar situation exists in several countries of North Africa where migration to the labour markets of Europe is adversely affecting agriculture.

There are also all the usual problems of inadequate inputs. Fertilizer and improved seed are either not available or too expensive for the subsistence farmer. He has inadequate investment capital, and cooperatives and sources of credit are usually undeveloped, and particularly difficult for semi-nomadic peoples.

There are also the social problems arising from the difficulty of effectively managing communal grazing land; and the reluctance to make investments of cash or labour in land which is held on a temporary or insecure lease. The infrastructure is usually less well developed in semi-arid regions because there has been both less demand for roads, water supplies, and marketing outlets, and also because the low level of productivity does not justify development of these facilities.

There are usually problems of distribution of seed, chemicals, and fertilizer, and these have recently been thrown into prominence by the inability to distribute food provided by aid agencies in countries suffering famine. While most governments have policies directed towards the improvement and development of agriculture, these are not always pursued as vigorously as they might be.

It has been suggested that in several countries in Africa the government policy of reducing the national herd to relieve the pressure on degrading grazing land is not happening because so many of the stock are in very large herds owned by powerful individuals. Another difficulty for governments is how to reconcile the political wish to hold down the cost of staple foods and yet at the same time pay a fair price to the farmers. At the level of international relations, it is alarming that there are so many political conflicts in semi-arid regions where the struggle against nature is bad enough without being made worse by further conflict.

METHOD OF WATER CONSERVATION

This section deals with methods to increase the amount of water stored in the soil profile by trapping or holding rain where it falls, or where there is some small movement as surface run-off.

There is no simple way of classifying methods of water conservation. One suggestion is to do it by comparing rainfall with crop requirements, giving three conditions:

i. Where precipitation is less than crop requirements; here the strategy includes land treatment to increase run-off onto cropped areas, fallowing for water conservation, and the use of drought- tolerant crops with suitable management practices.

ii. Where precipitation is equal to crop requirements; here the strategy is local conservation of precipitation, maximizing storage within the soil profile, and storage of excess run-off for subsequent use.

iii. Where precipitation is in excess of crop requirements; in this case the strategies are to reduce rainfall erosion, to drain surplus run-off and store it for subsequent use.

The weakness of this approach is that the main feature of rainfall in semi-arid regions is that it is very erratic and completely unpredictable. There can be wide variations of moisture shortage and surplus, both within and between seasons. A drought year whose total rain is well below the long-term average may still include periods of excessive rain and flooding, while a high rainfall season may include periods of drought.

This makes the choice of method difficult, because the desired objective may change from one season to another. In a dry area it may be sensible to increase surface storage to improve crop yield in most years, but in a wet year this could cause waterlogging and reduce the yield. On the other hand, a drainage system may have the objective of increasing the run-off but also have the undesired effect of exaggerating the effect of a drought. It is therefore not practical either to classify methods according to average conditions, or to design strategies based on averages. The art or science of water management is to reduce the problems caused by non-average events of flood and drought.

Sometimes it may be passible to have dual purpose methods which can be changed mid-season, for example by opening up the ends of contour bunds to shed surplus water after a wet start to the season, or to block outlets for the opposite effect.

In addition to the vagaries of rainfall, there are a host of other variables, the soil, the land use, the farming system, and the social patterns. Throughout the semi-arid regions of the world there is a wide range of indigenous systems and methods, but experience has shown that they do not always transfer well from one set of conditions to another. In a good discussion of this subject, Pacey points out that apart from the technical differences, methods must also be compatible with local life styles, social systems, and patterns of administration. Transferring what appear to be simple techniques requires not only the dissemination of information but also the adaptation to local conditions. In this Bulletin, when presenting information on available methods, we have not tried to prescribe a detailed treatment for a particular problem, but rather to set up a large array of possible methods, from which the person on the spot can choose methods to test locally.

When choosing a method or technique we must be aware of the physical and personal difficulties. Because of the variable rainfall we must expect a low success rate. The odds will be slightly better for grass, trees, and shrubs than for grain crops which are more demanding in their moisture requirements. They need rain at the beginning of the growing season, then a number of storms big enough to generate run-off and at reasonably regular intervals, since run-off events too close together will be wasted if they exceed the storage.

The soil must therefore have sufficient soil moisture storage capacity to keep the crop growing between run-off events. There will often be difficulties in keeping stock away from improved areas. Wire fencing is expensive, live hedges need time and care to establish them, and thorn hedges need regular maintenance. Maintenance is also required on any structural works like banks or furrows.

Personal constraints are that people naturally do not like to put a lot of effort into schemes which have a low success rate, nor if their tenure is insecure. In dry areas the people are frequently mobile or partly nomadic and so may not live permanently near sites suitable for run-off schemes. There are many former schemes which are now abandoned, and knowing more about the reasons would help us plan new schemes.

Some Design Principles

The term "rainfall multipliers" describes methods where the run-off from an uncultivated part of the land is diverted on to a cultivated part, thus giving it the benefit of more water than it receives directly as rain, hence rainfall multiplier.

The object of designing such methods is to get the best ratio of area yielding run-off to the area receiving run-on. Because of the variations of soil and crop requirement, and above all the variability of rainfall, this can never be a straightforward mathematical calculation. A formula which can be used to provide a starting point for field trials is given by Finkel, together with some empirical values for Turkana in Kenya, to be used in the formula:

$$\text{Ratio} = \frac{\text{Catchment area}}{\text{Cultivated area}} = \frac{\text{Crop water requirement - Design rainfall}}{\text{Design rainfal x Run-off Coeft x Efficiency}}$$

— Crop water requirement is estimated from evaporation data with a crop factor applied;

— Design rainfall is probably less than long-term average rainfall in order to give a better crop in low rainfall years, while the excess can overflow in wet years;

— Run-off coefficient is the percentage of rainfall which becomes run-off;

— Efficiency reflects the difference between rainfall distribution and the water requirements of the crop.

Schemes with internal catchments require high energy inputs for surface modification and so tend to be more common on crop land and in areas with annual rainfall of 250 mm or more. The ratio of catchment area to cultivated area is likely to be from 1:1 to 5:1. Schemes using external catchments may have much higher ratios, for example up to 30:1 in the case of the run-off farming in the Negev desert. Such systems may be practical at levels of rainfall down to less than 100 mm, which is lower than is possible for in-field schemes for cropland.

In developed economies it is reasonable to look at the long-term returns from schemes which require investment, but this is less helpful in semi-arid areas. Even when long-term rainfall records are available, and probabilities can be calculated, subsistence farmers or nomadic pastora-lists work to very short-term plans. Several examples are reported from Kenya by Finkel when soundly planned schemes were abandoned after poor results in the first year to two.

The Effect of Scale

The size of catchment areas has a bearing on the yield of run-off. Under the same hydrological conditions, small areas may have up to 50 percent, whereas river basin run-off is only 5 percent. There are two reasons for this. Large catchment areas are more likely to have places where there is temporary storage of run-off in surface depressions. Also small catchment areas may yield run-off from short showers which do not produce surface run-off from larger areas. The hydrological explanation is that small areas have a shorter gathering time (i.e. time of concentration) and so give run-off from shorter storms (with higher intensities) than storms which need a longer gathering time to produce run-off from larger catchment areas. Shanan and Tadmor quote the following average run-off yield from 100 mm annual rainfall:

Less than	0.02 ha	10-30 mm/ha/year
	3-10 ha	4-10 " " "
	300-500 ha	1 " " "

Methods For Crop Land

In many cases it is not practical to use only hand labour, and animal-drawn implements are necessary, or tractors. Some of the methods show promise of increased crop yields so a major feature of experimental trials is the search for cheap, simple, low energy methods and machines.

Broad Bed and Furrow System (BBF)

The Broad Bed and Furrow system has been mainly developed at the International Crops Research Institute for the Semi-arid Tropics (ICRISAT) in India. A comprehensive research programme was carried out on-station for eight years before being taken for on-farm adaptive research at Tadthanapalle in Medac district. It is a modern version of the very old concept of encouraging controlled surface drainage by forming the soil surface into beds. In medieval times in Britain this was used for improving pastures and called "rigg and furrow"; it has also been used in North America and in Central Africa. A variation known as the camber-bed system was used in Kenya

The recommended ICRISAT system consists of broad beds about 100 cm wide separated by sunken furrows about 50 cm wide. The preferred slope along the furrow is between 0.4 and 0.8 percent on vertisols. Two, three, or four rows of crop can be grown on the broad bed, and the bed width and crop geometry can be varied to suit the cultivation and planting equipment.

In India the system has been used mainly on deep vertisols (heavy black clay soils sometimes called cotton soils); wide beds are used on a gentle grade and they are formed by ox-drawn wheeled tool carriers.

All these schemes have, in varying proportions, the following objectives:

i. to encourage moisture storage in the soil profile. Deep vertisols may have soil moisture storage up to 250 mm, which is sufficient to support plants through mid-season or late-season spells of drought. The possibility is also increased of double cropping by means of inter-cropping or sequential cropping. The large water storage capacity of the soil supports growth more easily during the subsequent dry but cooler post-rainy season.

ii. to dispose safely of surplus surface run-off without causing erosion.

iii. to provide a better drained and more easily cultivated soil in the beds. There is only a narrow range of moisture conditions during which the soil can be efficiently tilled or planted, and timeliness is a key factor. Only about 20 percent of the deep vertisols in India are cropped during the rains, mainly because of poor workability when wet. The situation is similar in Ethiopia. If a crop can be established during the early rains, the profile is usually near saturation only for short periods during the latter half of the season, water is more efficiently utilized, and there is less need for run-off collection and storage. The possibility is also increased of double cropping by means of inter-cropping or sequential cropping. Tillage of the raised beds may be possible before the rains, introducing the possibility of dry seeding ahead of the rains in areas where the start of the rainy season is fairly reliable, and there is a good chance of follow-up rains to ensure the establishment of the germinating crop. The difficulty of preparing a seed bed during the dry season in these hard clay soils has been greatly improved by the use of broad beds and animal-drawn equipment.

iv. the possibility of the re-use of run-off stored in small. Small amounts of life-saving irrigation applications can be very effective in dry spells during the rains, particularly on soils with lower storage capacity than the deep vertisols.

The BBF system is particularly suitable for the vertisols. The technique works best on deep black soils in areas with dependable rainfall averaging 750 mm or more. It has not been as productive in areas of less dependable rainfall, or on alfisols or shallower black soils - although in the latter cases more productivity is achieved than with traditional farming methods. Other methods, with more emphasis on storage and irrigation within a package which includes BBF, are more likely to be viable for the alfisols. It is also stressed throughout the ICRISAT research that the BBF system should not be considered in isolation, but only as part of an improved farming systems package.

An important component of the system is an ox-drawn wheeled tool bar, which can be used with ridgers to form the raised bed and also later for carrying precision seeders or planters. The tool carrier is thus used for the initial forming of the beds, the subsequent annual reshaping, and for all tillage, planting, and inter-row cultivation.

Ridging and Tied Ridging

This method is also known in the USA as furrow blocking, furrow damming, furrow diking, and basin listing. The principle is to increase surface storage by first making ridges and furrows, then damming the furrows with small mounds, or ties. Graded ridges alone will usually lead to an increase of surface run-off compared with flat planting, while tied ridges will decrease the run-off and increase the storage. In different seasons either of these two effects may be preferable. And so the design question is when to go for drainage and when for storage. An aid to designing what is best in the long-term by modelling has been developed, but designing the system which will give the best effect in the long term will not prevent the system giving the opposite effect from that which is required in a particular season. The possibility has been suggested by Ahn of hedging the bet by tying alternate furrows. This is not a very satisfactory solution. One argument is that this would reduce the amount of damage by too much run-off or too much retention, but one could equally well argue that it will ensure that half of the land will always be under the wrong treatment.

An interesting possibility is to use tied ridging in connection with sprinkler irrigation to allow higher application rates at low pressure. Tied ridging is usually associated with mechanized farming. There have been some attempts at achieving it with ox-drawn implements, but the system really needs high draught for speed and precision, which is required if the ridges are to be re-ridged or split in subsequent years.

Either ridging alone or tied ridging has occasionally been practised using hand labour, but the high labour requirement usually makes this unpopular with subsistence farmers. Anyway hand-made ridges are usually less efficient. They are more likely to depart from a true contour and to have variations in the height of the ridge, both of which will increase the risk of overtopping..

Varying evidence of the results of tied ridging

There is an extensive literature reporting trials of tied ridging in many countries. A few of the reports indicate problems and failures but the great majority claim such outstanding success for the system that one wonders why the system has not been more widely used. A possible solution to this conundrum may be that the inconsistency, and unreliability of good results, prevents it becoming more widely adopted.

The experience in Africa has been well summarized by El-Swaify et al. as follows:

> "Under certain circumstances the system has been beneficial not only for reducing run-off and soil loss, but also for increasing crop yield. However, during high rainfall years or in years when relatively long periods within the rainy season are very wet, significantly lower yields were reported from systems with tied ridges than from graded systems which disallowed surface ponding ef water. Under such conditions tied ridging enhanced waterlogging, developed anaerobic conditions in the rooting zone, exçessive fertilizer leaching, and water table rise in lower slope areas."

In other countries there are conflicting reports, with a majority of successes. McCartney et al. reported that tied ridging in Tanzania gave higher maize yields not only in low but in high rainfall years as well. However, reports of success are more common in low rainfall years, for example Njihia reported from Katumani in Kenya that tied ridging resulted in the production of a crop of maize in low rainfall years when flat-planted crops gave no yield. Jones goes on to say:

> "Similarly on a sandy soil at Lusitu in the Zambesi valley, tied ridges increased mean crop yields (maize, sorghum, and millet) over those on flat land by 168, 159 and 16 percent under seasonal rainfalls of 587, 623, and 724 mm ; and on vertisols at Big Bend, Swaziland, mean increases for maize, cotton, and sorghum were 64 percent in a year of 508 mm, and 308 percent in a year of 310 mm. Clearly responses can be dramatic, but recent work under very harsh conditions in Botswana has shown that there may be also negative effects. Higher soil temperatures within the ridge can be detrimental to seed germination, and where showers are light the penetration of moisture into the soil may be shallower than that in the flat soil."

Another report from Botswana indicates a positive response to sorghum yields from tied ridging in both a very dry season (1972/73) and a very much wetter one (1973/74).

In the Negev desert in Israel, Rawitz et al. reported that tied ridging greatly increased the efficiency of a fallow before cotton, but that when the system was tried for wheat crops it was less successful because of waterlogging in the basins. The use of wide beds with untied furrows was therefore preferred and gave increased yields, but the climate of the Negev is special since it has winter rainfall and a very low mean annual rainfall.

From India also, broad ridges gave better yields of sorghum than either flat planting or narrow ridges and a similar result was reported for sorghum and castor in Gujarat by Brahmbatt and Patel. In Texas, USA, Clark and Jones reported substantial increases in sorghum yield from 1420 kg/ha to 1650 kg/ha for tied ridging compared with flat planting. In another American study in Texas diked furrows were found to be effective at holding and absorbing high intensity storms. Similar successful applications of tied ridging in the Great Plains of the USA are also reported by Stuart et al., and again one is inclined to wonder why, in the face of so much positive experimental evidence, one does not find the whole of Texas and the whole of Tanzania covered with tied ridging. One possible reason for the low adoption of the system is the temperature effect reported by Jones. Another is the danger of waterlogging in wet years. This can be worsened when there is compaction of the furrow bottom by tractor traffic, and one must also consider whether the machinery and implements are as efficient as the reports would suggest.

Conditions where tied ridging is suitable

There is a danger of soil erosion if the ridges are overtopped and break so that the water temporarily stored in the depressions is suddenly released. This will not happen if the combination of surface storage plus the amount which infiltrates into the soil surface is less than the storm rainfall. This implies a high value of soil storage, usually deep soils with good infiltration and permeability. In some systems the infiltration is increased either by mulching in the furrow bottoms or by subsoiling or cultivating.

Three safety back-ups are required to minimize the risk of damage by erosion:

— the furrows should be on a gentle grade to assist run-off if the ties fail;
— the ties should be lower in height than the ridges so that the ties fail along the furrows before the ridges fail down the slope;
— there should be a back-up system of conventional graded channel terraces to prevent damage if the ridges do overtop or fail.

Implements for building tied ridges

Making an implement to form the ridges is straightforward, it is interrupting the ridging process to leave a tie that is difficult. Possibilities are:

— intermittent lifting by hand if the ridger is pulled by tractor or by oxen;
— automatic lifting devices based on an eccentric wheel;
— intermittent hydraulic lift either manual or triggered by rotation of tractor wheels.

The design of implements in the USA is described by Lyle and Dixon, and in the USSR by Tregubov. An earlier review by Boa is useful but out of print.

Simpler systems for non-mechanized farming

The construction of tied ridges is mainly associated with mechanized farming because of the high labour requirement, but it is quite possible to do it by hand labour.

Conservation Bench Terraces (CBT)

This is another type of rainfall multiplier, using part of the land surface as a catchment to provide additional run-off onto level terraces on which crops are grown. The method is particularly appropriate for largescale mechanized farming such as the wheat/sorghum farmlands of the southwest of the USA, where the method was pioneered by Austin W. Zingg in 1955. Experimental results were first reported in 1959 and after detailed studies at Bushlands Experiment Station in Texas, and Hayes in Kansas, a technical and economic evaluation was made by Hauser and Cox. This led to extensive trials in the six western States with low rainfall, and these were reported in detail. CBTs were compared with the conventional practice of level terraces (i.e. level along the length, but the original slope is left between terraces), and also with all over bench terracing. There is a soil erosion hazard during high-intensity summer storms, but CBTs were as effective at controlling erosion as the other two practices, and more effective at reducing the overall run-off. The data from these trials provided general guidelines on the method, but standard designs should be avoided because of the wide variation in conditions of soil, rainfall and farming system. The best way of applying the system in a particular situation should always be investigated locally.

The features required for CBTs are:

— Gentle slopes of 0.5 - 1.5 percent are most suitable although the system has been used up to 6 percent in North Dakota. As with all terracing, a steeper slope requires more earth moving.

— A deep soil is required, both to provide sufficient soil moisture storage, and also to lessen the effect of cutting during the construction of the terraces. Several USA workers report a reduction of yield from the disturbance of the soil during terracing for up to six years. Some even went to the length of stockpiling the topsoil and replacing it after levelling. Good permeability is also required so that the contained flood water can be absorbed quickly.

— Smooth slopes are an advantage where large mechanized farming can be made more convenient by constructing all the terraces parallel and of equal width.

— Precise levelling of the bench terrace is important to ensure uniform build up of soil moisture.

— If there is a risk that run-off from the catchment area will be greater than can be absorbed and stored on the terrace, there must be outlets at the ends of the terrace, which can discharge into grass waterways or other safe disposals. As with tied ridging, if there is going to be surface run-off it must be along the terrace, not over the edge of the terrace and down the slope with consequent risk of erosion. For CBTs a common method is to make the level of the outlet at the end of the terrace one half of the height of the bund on the edge of the terrace, ensuring a freeboard of half the bund height.

— This is a high-cost system which has to be paid for by high yields and so a high level of inputs is required, particularly seed, fertilizer, and crop management. Adequate levels of fertilization are required to avoid this being the constraint after the moisture limitation has been removed.

— The main design factors are the width of the levelled terrace and the ratio of the area of the terrace to the catchment area. The width of terrace is governed mainly by the machinery to be used. As the width of planters and drills continues to increase, so does the need for wider terraces, but this is restricted by the slope and the depth of soil. Typical widths are from 10 metres on land of 5 - 6 percent, to 30 metres at 2 percent, and 50 metres or more at one percent. Jones reports mini-terraces 9 m wide with a 1:1 ratio and different rotations on the catchment and the terrace. Total yield was doubled on the terraces.

— The most usual practice is for the bench terrace to be level along its length, but there have been a few trials with the terraces on a gentle gradient of 1:400 to encourage safe disposal of excessive run-off.

— The ratio of catchment to terrace may be from 0:1 (in which case the whole land surface is converted to level bench terraces) but typical values are 1:1 or 2:1. In general, the lower the mean annual rainfall the larger will be the required catchment area to provide sufficient moisture for the crop. Another consideration will be the type of rainfall since run-off will only be produced from storms where the intensity exceeds the infiltration on thecatchment area. The treatment of the catchment area will also effect the amount of run-off. In very dry areas the catchment may be left uncultivated to promote run-off, while in the south western USA the usual practice is to have continuous cropping on the bench terrace with a rotation on the catchment area. A common rotation on the catchment is wheat/sorghum/fallow, giving two crops in three years.

Results of the American experience show that the method is suitable and economic for the high-input large-scale mechanized farming of the region. The costs of land preparation can be expected to be recovered in about 10 years. Complete bench terracing of the whole land surface sometimes gives the greatest cash return since the whole of the land is cropped, but the cost of terracing is much higher. With a ratio of 1 terrace:2 catchment the terracing costs are only one third, and in several studies this was the most economic system.

Contour Furrows

These are variations on the theme of surface manipulation which require less soil movement than conservation bench terracing, and are more likely to be used by small farmers, or in lower rainfall areas. The cropping is usually intermittent on strips or in rows, with the catchment area left fallow. The principle is the same as Conservation bench terraces, that is to collect run-off from the catchment to improve soil moisture on the cropped area. Studies were made of the soil moisture profile with a catchment ratio of 2:1; a satisfactory sorghum crop was grown on only 270 mm of rainfall. It was estimated that run-off from the catchment was 30 percent giving 166 mm of run-on, and 432 mm available to the plants:

$$\left(\frac{30}{100} \times 2 \times 270\right) + 270 = 432$$

On heavier soils, contour bunds may be less effective because of the lower infiltration. Studies on vertisols in India showed that yields were lower near the bunds, both upslope and down, as a result of waterlogging.

Contour bunds are also used in Ethiopia for a combination of soil conservation and water conservation. The bunds are built on level grade with ties in the basin. A stone wall is built on the lower side of the earth bund in an attempt to reduce damage if the basin is overtopped, but this has not been very successful, and it would be better to put the bunds on grade if it is likely that they will not be able to hold all the run-off.

If the contour furrows are not laid out precisely on the contour, or are built with some irregularities, there may be a danger of uneven depths of ponding behind the bank. This can be reduced by smaller bunds at right angles, but as with tied ridging, these bunds should be lower in height than the main ridges so that if there should be any overtopping it will be laterally along the contour and not over the bund and down the slope. Sometimes the emphasis is on the excavated furrow which collects water, so that in exceptional storms the run-off can overflow without damage.

The energy input for building the ridges is still considerable but several ox-drawn ridgers are available, and more expensive but more efficient tool bars. Even if the whole operation has to be done by hand labour this may be acceptable if it is replacing a traditional system which also has a high labour requirement. This was the situation in the Kenya example reported by Smith and Critchley, where the traditional method consisted of deep digging by hand to increase the infiltration and moisture storage.

Another approach to localized surface storage on part of the land is the use of small semi-circular catchments or larger rectangular or trapezoidal structures. The large number of separate semicircular or trapezoidal bunds spreads the risk of damage from overflow in exceptionally heavy rain, or if one bund fails. The semi-circles are usually constructed by hand, and are larger versions of range pitting which is used to promote vegetation on grazing land. In several semi-arid climates in Africa the farmers have shown more interest in using these techniques for crops than for range improvement. The trapezoidal bunds were developed by Finkel in Turkana, Kenya, and can be built by hand or machinery.

The trapezoidal bunds can be from 0.25-2 hectares, and so are more suitable for cultivation. They can be levelled within the bund to ensure more uniform spreading of the retained water, but this may not be a good idea if there is a concentration of fertility in the topsoil. One tip of the bund may be slightly lower to serve as an overflow spillway, which can be protected with simple stone paving if there is any danger of erosion. Both semi-circular and trapezoidal bunds can be staggered on alternate rows, so that the overflow from one row will run into the next downslope, and ensure that all are filled before there is any run-off from the field. Both types are best on gentle slopes, especially trapezoidal bunds which are best limited to slopes of less than 2 percent, because at steeper slopes the bunds have to be built up higher and so need a high labour input.

Where the whole of the soil surface is shaped, the ratio of run-off area to run-on area is governed by the configuration. An important advantage of small run-off producing areas compared to larger ones is that they will yield run-off from smaller rain storms At the Central Arid Zone Research Institute in India, using narrow strips with a ratio of run-off to run-on of 0.75 (catchment) to 1.0 (cropped), threshold values of rainfall for inducing run-off were between 3 and 15 mm for moist surfaces and 7 to 9 mm for dry surfaces. The possibility of increasing run-off through surface applications of bentonite or sediment from tanks was possible but unlikely to be economic. Another possibility which could have practical significance is changing the configuration to increase the slope of the run-off area. A system of creating furrows on sloping land using asymmetric ridgers was shown to be successful for maize production on an experimental basis in Mexico.

Low cross-ties can be used to help obtain a better distribution of water since they reduce accidental run-off if some of the lines have a fall. Fairbourn also showed that infiltration could be increased by burying crop residues in a narrow trench in the cultivated area; he called this vertical mulching but the extra complication of trenching and filling make this practice unlikely to be adopted on a wide scale. It bears some relation to a simpler variation in which crop residues are thrown into the bottom of furrows and the ridges are split onto the furrows to form a new ridge for cropping the following year. Subsoiling or deep ripping in the furrow may also be used to increase infiltration, particularly when there is a clay pan or a laterite layer.

Treatment of the run-off area with additives is unlikely to be economic, but it may be possible to take advantage of naturally occurring soil properties to increase the run-off. A case in point is Botswana where the sandy soils known as the 'hardveld' become hard and crusted after heavy rain. A method proposed there by Nilsson consists of forming permanent low ridges about a metre wide which are left to develop this hard crust, while ploughing, cultivating, and planting are restricted to narrow strips of 30 cm. This system has the advantage that after it has been established, no further maintenance is required.

The concept of strip tillage on the contour with ripping to encourage infiltration was tested in Tanzania by Macartney et al., and in Botswana, Willcocks showed that precision strip tillage was more effective than other cultivation methods, particularly if the strips could be consistently aligned along the previous crop rows.

In low rainfall areas a large ratio of catchment to cropped area is required, but it is not easy to design a system which will give the best result for all variations of annual rainfall. In some experiments in North America the ratio of 33:1 was tried, and with 190 mm of rain gave 530 mm of run-on to the farmed area, which was more than required and more than could be absorbed. The next year the ratio was reduced to 15:1, but the rain was greater at 246 mm and with a higher run-off coefficient gave 390 mm of run-on, about the right amount. Retaining the ratio of 15:1, the next year had 140 mm rain, giving 220 mm run-on which was not sufficient. The authors suggest that mathematical modelling is an appropriate tool to design the optimum ratio for varying conditions of soil and climate, but as we have discussed previously, any system is going to have widely varying degrees of success or failure from year to year. In a region of low winter rainfall, using saw-tooth ridges shaped by motor grader, Shanan and Tadmor found that the best ratio of catchment to planted strip could vary from 4-20. The very wide range suggests that trial and error may be as effective as mathematical modelling.

Water Spreading

Here again we have a wide variety of methods for different conditions. The soil type influences the method in several ways. Clay soils have low infiltration rates and high moisture storage capacity, so they are suitable for deep flooding with subsequent cropping. A deep soil can absorb larger amounts of water, while a shallow soil may need the provision of overflow outlets to avoid drowning the crop.

The climate too; affects the method. Mediterranean climates have winter rainfall and low evaporation which may make possible run-off farming with mean annual rainfall (MAR) down to 100 mm as in the Negev desert in Israel and in Tunisia. But with tropical summer rainfall and high evaporation, water harvesting is most likely to be useful in areas with more than 250 mm.

The probability of high-intensity rain influences the method. In the Negev most rain falls in light showers so the objective is to maximize the run-off. In tropical Africa the problem is to handle the sudden floods from violent summer thunderstorms.

If the rain and the growing period do not coincide, the object will be to store water for later use. If the rainfall and growing period are the same, the desired result will be several applications of water during the period of crop growth.

Natural Run-off

There are many examples of traditional use of naturally occurring run-off to augment rainfall in areas where the rain alone is not sufficient for growing crops. Some authors have described this as "exploiting micro-niches in the environment".

In North America the culture of many Indian tribes has depended for centuries on simple methods of floodwater farming. For example, the Navajo in Arizona use run-off from sandstone outcrops to water alluvial soil at the base of the hills. Good crops of maize, squash, and melons are produced where the annual rainfall is only 300-400 mm. This type of scheme usually does not use a large investment of labour to manage the water, but advantage may be taken of any favourable feature, such as a road which can act as a collector drain, and to some extent the run-off may be managed by leaving the upper part of the field unplanted, or increasing the storage by a simple impounding dam.

Several other tribes use floodwater farming extensively in the south-west of North America, for example it is the practice of the Hopi tribe on three quarters of their cultivated land, although the area so used today is only one third of what it was in former times.

Four situations where simple, unimproved run-off farming may be useful are:

— alluvial or colluvial deposits at the base of ridges or escarpments;

— alluvial fans where streams leave hill land and flow onto flatter plains;

— stream banks which are intermittently flooded naturally;

— where run-off collects in naturally occurring depressions, as the 'cuvettes' in West Africa.

In Africa these conditions are all exploited, for example a case in Turkana, Kenya, reported by Morgan and Hillman. In a mainly pastoral community with annual rainfall of less than 200 mm, small patches of sorghum are grown in all the first three situations. Local traditional practice and familiarity with local conditions may be useful when considering possible sites for the development of more ambitious run-off farming schemes, and this was the case in the Turkana example.

Where the rainfall is sufficient for general cropping (and this usually means more than about 500 mm), it may still be possible to make use of run-on areas by increasing the cropping intensity. In the use of conservation bench terraces, separate rotations are used for the run-off area and the run-on terrace. The same principle is applied to natural run-off in Morocco. When growing wheat or barley, a plant density of 70-90 plants/m^2 is used on land receiving rain only, but the plant density is increased to 90-120 plants/m^2 in depressions which collect run-on. This does entail an element of chance, described by Pacey as "the deliberate reduction of plant density results in some sacrifice of yield in seasons of ample rainfall, but it makes for greater certainty that there will be useful grain production in a drought year."

Two examples are reported by Kovda of the use of naturally occurring run-off in arid areas of the USSR. 'Kair' farming is the name given to cropping on flood terraces of the large rivers of central Asia where the soil moisture is partly the result of surface flooding, but also by lateral seepage from the river. Another example is 'khaki' farming in Turkmenistan where run-off from rain on higher mountain slopes inundates gentle slopes in the plains. The flooded fields are left to dry out and then cultivated and planted as soon as is practicable. A similar practice is reported from Ethiopia by Carr. In the Woito and lower Omo valleys, as soon as the annual flood recedes, the flood plains are cleared and planted to maize and sorghum, using heavy seeding rates which are thinned later.

Collected and Diverted Run-off

The previous section describes some uses of natural or unimproved run-off, while this Section is about schemes where there is some element of manipulation or management of the land or the run-off. The difference is not important, because some systems use only very simple works.

It is not important whether we consider run-off farming as a type of irrigation. In the case of terraced wadis and other water management schemes in valley bottoms, man sets out to manage both the land and the water. Farming valley bottoms is a very ancient practice and has been well documented in Tunisia and in the Negev desert in Israel. In the Matmatas of Southern Tunisia the system is called 'Gessours'. In an area with only 100-150 mm of rain, rows of olive trees are planted across the width of the valley bottom near a minor barrage, which slows down the floods, and result in deposition of silt with a terracing effect. Cereal crops are grown on each terrace upstream of the olive trees. This is similar to the terraced wadis in the southern Negev in Israel. In some cases these ancient schemes are partly cultivated today by the mainly pastoral Bedouin inhabitants.

Another example of carefully controlled water use in valley bottoms comes from Colorado. Level pans of one to three hectares are formed in broad valleys where slopes are less than 3 percent. The rainfall is 400 mm, sufficient for some rather unreliable cropping of grain or forage sorghum. The cropped land yields 5 to l0 percent of the rainfall as run-off from the heavy summer storms. The run-off from 150 ha of cropland is spread onto the first level pan, and when the depth reaches more than 100 mm the surplus spills over to the second pan and so on. During the season an additional 200 mm of water results in good crops on the pans, so that the installation costs can be repaid in three to five years. In most schemes involving flooding of crops, it is usual to provide outlets for emergency overflow, and possibly for draining as well, because growing crops appear to be less affected when there is a slow continuous passage of water than if the water is stationary. Also most crops only tolerate complete saturation for a short time.

Some of the ancient Negev valley floor cultivation systems used only spate run-off down the valley, but most of them depended also on the collection of run-off from the surrounding hills. A fascinating programme of research has been conducted for twenty years by the universities and research institutes of Israel and has been splendidly recorded. The surface management to induce run-off included collecting and piling all the loose surface stones into mounds, and shaping the surface into ridges which are very similar to the roaded catchments used nowadays in Australia.

The collection systems in the Negev show that the hydrologic principles were fully understood. Stone-lined conduits were used to take the water quickly down to the farms with minimum loss by infiltration and evaporation, and the catchments for these drains were long and thin so that rain was collected even from the low intensity winter rainfall common in this area. The distribution systems were also simple in construction but sophisticated in concept, and allowed the water to be transferred and diverted around the farm quickly and efficiently with minimum labour requirements.

As a result of the successful studies in the Negev desert, trials of run-off farming were carried out in the Khost plain, in Paktia province of Afghanistan. The terrain is similar to the Negev; stony barren hills where run-off is collected in channels and led down to the plain. Level fields, surrounded by an earth bank, are laid out in shallow steps so the water is passed down-slope from field to field. After the soil profile has been saturated a cereal crop is grown on a combination of subsequent irrigations if there is rain, and the stored moisture in dry periods.

Another ancient collecting system called 'meskats' comes from the Sousse region of Tunisia with a rainfall of about 300 mm. Olives are grown on the better soils at the bottom of slopes. On the upper slopes, the soils have a tendancy to surface crusting and yield run-off into long thin catchment areas feeding the water down in channels to the olive groves.

These methods, developed for regions with winter rainfall, may not be suitable for the semi-arid tropics. Summer rainfall is more intense, so in general the catchment area required will be smaller, and provision for the safe passage of exceptionally high rates of run-off must be included. Although not yet adopted to any great extent by the local farmers, the national policy of land allocation to what has traditionally been a mobile pastoral community, will probably lead to a stronger motivation for regular cultivation on permanent plots, which will not be possible without some form of water harvesting. Ratios of catchment area to cultivated land of up to 5:1 are expected to be suitable. For the system to work well, the cultivated area should be levelled so that there is little or no lateral slope along each terrace, and only a gentle slope across the terrace to avoid erosive flows. Stone-packed spillways built into the earth bund allow surplus water to pass progressively downslope to lower terraces. The danger of damage from excessive floods should be avoided, not by trying to control the water once it has reached the cultivated area, but through the provision of overflow by-passes set into the bank of the collection drain. There is no doubt that such a system can be made to work on an experimental or demonstration basis, and the question is whether the pressures of social change will make it sufficiently attractive to people unaccustomed to a cropping system which requires this degree of precision and maintenance.

Another form of water spreading in USSR known as 'liman' irrigation is described by Kovda. This consists of what are described as contour ramparts across the slope, perhaps an apt description since they are large structures up to two metres high and four or five metres wide (similar to the 'murundum' water conservation structures used in Parana State in Brazil). Sometimes drain holes are used to pass the water through to lower levels, although an open spillway would appear to offer a simpler and more effective way of controlling the water movement.

The diversion of run-off water onto prepared level terraces is another ancient and widely used method. This is the basis of growing irrigated paddy in humid areas where water supplies are assured, and the same principles may be used to obtain benefit from intermittent water supplies in semi-arid regions. For example, in the mountainous dry areas of northern Pakistan, in the district of Gilgit and the Hunza valley, agriculture is entirely dependent upon the diversion of springs, streams, and melt waters from snow and glaciers, onto terraced alluvial outwash fans which are accurately called oases. Some amazing construction feats have been used to carry the 'khuls' across unstable scree slopes, and laboriously cut by hand out of rock faces.

Another interesting example of diversion onto terraces comes from the Manakhah region of the Yemen Arab Republic. In an area of annual rainfall of 400-600 mm, level bench terraces are constructed on steep slopes up the hillsides, with rock risers 2-3 m high and the terraced fields 2-10 m wide. Run-off from the uncultivated and unterraced upland is carried to the terraces in diagonal ditches reminiscent of those in the Negev. Stone-lined canals carry the water progressively down through the terrace system, sometimes with the canals running underneath the terraces. This sophisticated and efficient system is presently falling into decay because of labour migration to the oil producing countries of the Middle East. A similar system in the Yemen Arab Republic, called 'sayl' irrigation, is the diversion of flood water out of wadis onto levelled terraces.

Inundation Methods

This section describes systems where flood waters are impounded and retained long enough to saturate the soil so that a crop can be grown on moisture stored in the soil.

Simple systems on a small scale are used in the Sudan. On gently sloping land embankments known as 'teras' collect and hold surface run-off, and after the water has soaked into the soil a short duration millet is planted and matures in eighty days. In the semi-arid north-east of Brazil, the government research organization CPATSA in Petrolina has studied a similar method, and produced designs for the optimum shape and size of the bunds. There are also recommended cropping methods to make maximum use of the stored moisture, with variations in the cropping pattern depending upon the amount of rainfall and the amount of run-off stored. The inundated areas are known as 'vazantes'.

In India the use of contour bunds to retain run-off has a mixed story of success and failure. When used mainly as a soil and water conservation measure in areas of reasonable rainfall (750 mm - 1250 mm) on medium-deep vertisols, it was found that the disadvantage of waterlogging in the vicinity of the bund both uphill and downhill exceeded the advantage of increased cropping from the stored moisture in a dry season. But in areas of lower rainfall the method has been more successful, for example in the Siwana district of western Rajasthan. With a rainfall of 250 mm or more, low contour bunds only 0.3 m high divide the land into strips which are progressively inundated down the slope with flooding to a depth of 0.2 m during the monsoon. When the rain ceases, the stored water infiltrates and a winter crop is sown.

In semi-arid areas of India, large-scale inundation schemes have been used for hundred of years, some with sophisticated forms of water control and land management. In Bihar and Uttar Pradesh there are thousands of 'ahars' covering a total of more than 800 000 ha. Low earth bunds are built to retain run-off during the monsoon and when used on very gentle slopes, sometimes as low as 0.01 percent, water stored to a depth of one metre will throw back several kilometres, so large areas of land are covered. Most ahars flood less than 500 ha but others are as much as 4000 ha. The volume of water stored is less important than the area of land which is submerged. The soil must have sufficient depth and moisture-holding capacity to store enough moisture to carry a five month crop which is grown in 'rabi', the dry winter season, after the end of the summer monsoon season, 'kharif'.

The retaining bank is usually not more than 3 m high and may extend several kilometres on the contour, with some examples up to 10 km. There is usually a wide waste-weir to serve as emergency spillway, usually stone pitched with a crest level one metre below the top of the bund to give one metre freeboard for wave action. The emergency spillway is required partly because of the danger of a heavy storm occurring when the tank is nearly full, and also because the ahars are usually built in a series down the slope and the failure of any one would imperil those lower down the slope.

It is also usual to provide metal sluice-gates set in concrete structures to allow for quick emptying when necessary at the end of the rains. The water released may be collected in the next ahar down the slope. Sometimes pipes are laid through the bund if some of the discharged water is to be used for irrigation.

The main crop is winter wheat planted as soon as the flooded land has dried out sufficiently, and then grown out on the stored moisture. Sometimes a subsidiary crop is taken during the summer monsoon while the land is flooded, using a variety of rice known as 'floating rice' because it can grow up through the standing water. This technique of taking one crop of floating rice followed by a second crop after the water has receded is also practised in Thailand.

A secondary advantage of this system is that the soils in these semi-arid areas often have a tendency to salinity, and this is controlled by the regular leaching by the inundation water. The infiltration of considerable amounts of water may also have the effect of raising the general level of the water table near to the ahars, with improvement of the supply in shallow wells.

Another form of inunadation farming is used in the Jaisalmer district of Rajasthan. Here the topography is more broken, and the method is to build an earth bund across the valley plain to catch and store run-off and silt from the surrounding barren hills. These are called submergence tanks or 'khadins'. In the district of Jaisalmer there are more than 500 such tanks with an inundated area of more than 12 000 ha. A catchment ratio of at least 15:1 is usual and the tanks are designed to fill from rainfall of 75-100 mm, in an area where the annual rainfall is 165 mm with great variability. The main construction features are the same as the ahars, that is a wide emergency spillway and sluices to release the water. The depth of water stored varies from 0.5 to 1.25 m and it usually disappears from seepage and evaporation by early November, when the winter crop is sown, either winter wheat or the grain legume Cicer arietinum, (Bengal gram or chickpea). Similar systems of inundatiion farming Pakistan are known as 'sailabas' and 'kuskabas'.

Secondary benefits of most inundation systems are the leaching of salinity by the stored water and the improvement to the yield of wells downstream.

Flood Diversion

his section is concerned with diversion and spreading of floods and spate flows.

Diversion of flood water from its channel usually involves some form of structure, a barrage or weir to divert the water. For small schemes, simple diversions may be constructed each year using stones and boulders, perhaps in wire netting, or poles and brushwood where these are available. It may be sensible to expend only limited effort on building such structures where The floods are kely to damage the conveyance channel unless it is provided with safety devices to spill water picked up in large floods. To avoid these problems a method has been widely used in India for centuries, the inundation canals, or 'pynes'. A canal is excavated in the river bank so that the level of the canal bed is considerably higher than the bed level of the river. During low flows the canal is dry, but when the flood level rises to that of the diversion canal it starts to flow. There is no need for any structure in the stream or river channel. This system is widely used in the Sind province of Pakistan to collect water from the Indus river and its tributaries during the flood period from April or May to September. The grade of the distribution canals is slightly flatter than that of the river bed to increase the command of the diverted flood water. This also means that there is heavy deposition of sediment particularly at the beginning of the canal, and regular clearing is essential. Some control is also required of the diverted flood water and this is usually done by a regulator several kilometres from the river. Subsequent control and distribution of the water is the same as for any surface irrigation scheme, except that the flow is ephemeral and used for heavy inundation so less precise control is required.

The other possibility is to divert the flood water out of its channel by raising the water level through some form of weir or barrage. These may be temporary and expendable in simple schemes, or permanent structures. The problems to be overcome by all flood diversion schemes are:

— damage to the diversion works during flash floods;
— control and spreading of the diverted flood water;
— deposition of sediment carried within the flood water.

In some cases the last one may be considered beneficial, as in the flood irrigation in the valley of the Nile, and flood irrigation in China from silt-laden rivers, where this is known as 'warping' and considered to be part of the process of maintaining fertility. The mica-laden melt waters from the glaciers of the Himalaya are used to build up the physical volume of soil by spreading the water onto gravel beds. On the other hand, too much deposition of sediment can lead to undesirable change of soil texture, or to reduced yield of grassland.

Surface Drainage

The main problem in semi-arid climates is shortage of water, but there are still occasions when at particular times or on special soil conditions the problem is too much water, and some form of drainage is required.

A soil where some form of water control can be important is the vertisols, heavy clay soils, usually dark, also called cracking black clay soils, cotton soils, and self-mulching clays. They are also known colloquially as 'one-day soils' because they are difficult to cultivate when dry and hard, and also unworkable when saturated and sticky, but they do have a short window while drying out when the moisture is just right for cultivation. The physical properties causing this are a high clay content, with the clay mineral being montmorillonite or illite, which, because of their 2:1 lattice structure, expand when wet, and shrink and crack when dry. A characteristic surface feature is the development of an uneven surface with raised mounds and shallow depressions caused by repeated cycles of wetting and drying and known in Australia as 'gilgai'.

The two main kinds of formation are the large uniform areas derived in situ from extensive volcanic flows of basalt, or smaller localized alluvial deposits in valley bottoms as in Ethiopia, Zimbabwe, and Zambia. In both situations, vertisols have considerable economic significance because they are a land resource with under-utilized potential owing to the difficulty of handling them.

One technical solution is the installation of subsurface drainage, and this was shown to be effective more than twenty years ago in Zimbabwe. Pipes of concrete, fired clay, and bamboo were all effective but uneconomic because of the cost of materials and transport. Today, the most appropriate method would be the installation of plastic tube drains by the trenchless method. The techology is available, but a high return would be required to justify the installation costs. Mole drains would be cheaper but the extremes of wet and dry conditions would lead to a short life.

In India, surface drainage can lead to the opportunity for early dry-sowing just before the start of the monsoon where the onset is fairly predictable. This allows double cropping instead of the traditional practice of the vertisols lying fallow during the monsoon, and only being cropped in winter on the stored moisture. In Ethiopia the occurrence of vertisols is in smaller areas in valley bottoms, which remain either uncropped altogether or only cropped towards the end of the rainy season. Improved ease of cultivation, through surface drainage by raised bed and furrow or by shallow open drains, could lead both to improved production from the areas now cultivated, and also to a large increase in the area used for crop production.

Vertisols offer the most opportunity for the use of surface drainage, but there are other situations where surface drainage may be useful. The BBF system has also been developed on alfisols in north eastern Brazil.and a variation on the same theme called ridge-and-furrow has been used on quite light soils in Zimbabwe and Kenya. In the last two cases, shallow soils become waterlogged because of a limited moisture storage in the profile, and a simple surface drainage system can lead to better opportunities for cropping, or an extended season for grazing. An interesting but specialized need for surface drainage comes from Ethiopia in the growing of teff which is the main food grain

Another example of surface drainage to solve a particular problem comes from Portugal where the situation is temporary waterlogging at the time of preparing or planting winter cereals. The method is called 'pnudivales' which are small channels constructed after seed-bed preparation and before seeding. They are made with three passes of a tractor. First with a mouldboard plough to create a furrow, followed by two passes with a mounted ridger to produce a shallow open drain between 3 and 4 m wide and 0.25 m deep at the centre. This allows subsequent drilling, cultivating, and harvesting in straight lines ignoring the open channels. These are set out on a gentle gradient of 0.6 to 0.8 percent.

Snow, Dew And Mist

In high latitudes, some or all of the precipitation may fall as snow, particularly in the northern hemisphere, and in semi-arid regions there may be the possibility of increasing the effectiveness of the precipitation by management of the snowfalls.

Trapping snow during the winter to increase available moisture in the short summer is a regular practice on the semi-arid steppes of Eastern USSR in Kazakhstan. Some of the first snow is caught by the stubble of the preceding crop, or, in the case of summer fallow, by low lines of mustard planted for the purpose. As soon as 50 or 100 mm of snow has accumulated it is pushed into ridges by heavy-duty tractors with a V-shaped bulldozer on the front and an inward sloping half grader blade on either side, so that the snow is pushed into two windrows or snow ridges one on either side of the tractor. A typical design is 6 m between centres of the ridges with 3 m of untouched snow between. The ridges are up to 1 m high and 3 m wide, and serve as wind breaks to catch more snow between the ridges. In a year of very heavy snow the total snow pack between the ridges can be up to 600 mm but this is more than desirable since the melt water causes erosion problems; 400 mm is normal and more desirable.

The conservation bench terraces have the main purpose to hold on the level terraces the run-off from catchment areas on the uphill side. In the two most northern states of the USA where the method was tested, it was found that they also increased the amount of trapped snow, partly as a result of the terraces acting as a physical barrier, and also because of the stubble of the previous summer crop grown on the terraces. Two other methods of snow trapping have also been practised in Montana. Neff reports that contour furrows caught snow equivalent to an additional 22 mm of rain, and also reduced the run-off from winter rain and the snow melt in the spring. In eastern Montana, Saulmon used a standard snow fence (1.25 m high with vertical wooden slats) to trap snow on small catchments which provide surface run-off for small stock watering ponds. The additional depth of snow gave an increase of 100 mm of surface run-off.

The extensive literature on dew as a source of water contains many qualitative opinions about its importance or lack of it but few quantitative scientific studies. The most detailed study in semi-arid areas is that in the Negev by Evanari et al.. They conclude that the annual total of dew may be significant in comparison with the annual rainfall. An example is a range of 25-35 mm from dew at Avdat where the mean rainfall is 150 mm. The dew formation is considerably more reliable than the rainfall, so that it is possible, as in 1962/63, for the annual dew formation at 28.4 mm to exceed the rainfall at 25.6 mm. This is very much greater than the amounts of dew formation precisely measured in England by Monteith who found that the average dewfall was from 2-5 mm. However, the desert dew is made up by a large number of very small amounts and not of any consequence as a water supply which can be collected, nor for any higher forms of desert plants, although it is used to some extent by the lower forms like lichens and xerophytic algae.

The Negev studies also showed that the piles of stones sometimes previously assumed to be dew collectors for the production of grape vines do not acquire enough dew on the surface of the stones for it to drip off, and that the mounds are in fact for the purpose of increasing surface run-off. There are also reports from China of a long history of growing melons in soil beds covered with a layer of gravel 100-150 mm thick and known as 'gravel fields for melon'. Condensation and dew formation dripping down to the soil is supposed to be the purpose of the gravel but its effect may perhaps be to maintain an even soil temperature or to reduce evaporation.

There are a number of examples of artificial collectors, such as plastic sheets being used to collect enough dew to keep alive forest seedlings in the desert in Israel, and another use of plastic sheets to produce emergency supplies of drinking water in Australia, but these do not have any agricultural significance.

There are also reports of aerial wells which supply drinking water in many dry areas, some with piped supplies, but the studies in the Negev would suggest that the process is more likely to have been the interception and condensation of fog or mist, or direct run-off, than dew formation. The story of dew formation and mist condensation is interesting, and full of contradictions, but the conclusion has to be that this is not a significant source for water for agriculture.

References

Boers T.M. and Ben-Asher J. *Harvesting water in the desert.* In: Annual Report 1979, International Institute for Land Reclamation and Improvement (ILRI) Wageningen, Netherlands. 1979

Cannell G.H. (ed). *Proceedings of an International Symposium on Rainfed Agriculture in Semi-Arid Regions,* Riverside, California, April 1977. 1977

Matlock W.G. and Dutt G.R. *A Primer on Water Harvesting and Run-off Farming.* 2nd Ed. Agric. Eng. Dept., College of Agriculture, University of Arizona. 1986

National Academy of Sciences. *More Water for Arid Lands. Board of Science and Technology for International Development,* Commission on International Relations. NAS, Washington DC. 1974

Pacey A, with Cullis A. *Rainwater Harvesting - the Collection of Rainfall and Run-off in Rural Areas.* Intermediate Technology Publications, London. 1986

UNEP. *Rain and Stormwater Harvesting in Rural Areas.* Ed. UNEP, Tycooly International, Dublin. 1983.

7

On-Farm Water Conservation

About 70% of the world's fresh water is being using in agriculture every year. This is why farming should focus on abundant water conservation. Farm water, also known as agricultural water, is water committed for use in the production of food and fibre. On average, 70 per cent of the fresh water withdrawn from rivers and groundwater is used to produce food and other agricultural products. Farm water may include water used in the irrigation of crops, water used to leach harmful salts from agricultural fields and water used for environmental management.

At this time, there were fewer than half the current number of people on the planet. People were not as wealthy as today, consumed fewer calories and ate less meat, so less water was needed to produce their food. They required a third of the volume of water we presently take from rivers. Today, the competition for water resources is much more intense. This is because there are now nearly seven billion people on the planet, their consumption of water-thirsty meat and vegetables is rising, and there is increasing competition for water from industry, urbanisation and biofuel crops. To avoid a global water crisis, farmers will have to strive to increase productivity to meet growing demands for food, while industry and cities find ways to use water more efficiently.

Successful agriculture is dependent upon farmers having sufficient access to water. However, water scarcity is already a critical constraint to farming in many parts of the world. Physical water scarcity is where there is not enough water to meet all demands, including that needed for ecosystems to function effectively. Arid regions frequently suffer from physical water scarcity. It also occurs where water seems abundant but where resources are over-committed. This can happen where there is overdevelopment of hydraulic infrastructure, usually for irrigation. Symptoms of physical water scarcity include environmental degradation and declining groundwater.

Economic scarcity, meanwhile, is caused by a lack of investment in water or insufficient human capacity to satisfy the demand for water. Symptoms of economic water scarcity include a lack of infrastructure, with people often having to fetch water from rivers for domestic and agricultural uses. Some 2.8 billion people currently live in water-scarce areas.

According to the researched data, water from major aquifers in the underground is being withdrawn faster in these days as a result these aquifer cannot be recharged by rainfall. Relevant study said that the farmer should use various water conservation methods to protect forests and wetlands, encouraging them to use water more efficiently and smartly. Some efficient on-farm water conservation methods are.

— *Irrigation scheduling*: it is so much important to decide when and how much water needs to apply to a field as it has a significant impact on irrigation efficiency and water-use efficiency as well as the amount of water used by the crop. In order to determine when to irrigate a number of different and innovative scheduling systems were developed. These different scheduling systems are proven, reducing the use of total amount of water to a certain area of a field and improving yield.

— *Tailwater return systems*: surface irrigation requires a certain amount of water be drained off as tailwater to provide sufficient water to the low end of a field. The tailwater return systems catch and pump this runoff to the top of the field.

— *Irrigation system improvement*: This method is an advanced method that uses computer software and hardware to apply the water properly to the field, minimizing water losses tremendously. The method increases the water distribution uniformly and reduces applied water. Depending of the soil type and applications, it may be very efficient.

Farm Water Management

Water, or the control of water, affects most crop production activities.

Sufficient water must be present in the rootzone for germination, evapotranspiration, nutrient absorption by roots, root growth, and soil microbiological and chemical processes that aid in the decomposition of organic matter and the mineralisation of nutrients. These factors are all necessary for sustaining crop growth on a particular field. At the same time, the rootzone must be sufficiently dry to ensure adequate aeration and root growth. The rootzone must also be dry enough to allow field access for performing cultural practice activities such as planting, cultivating, fertilisation, pesticide and herbicide applications, and harvesting.

Water movement through the soil is necessary in order to leach excess salts from the rootzone and so enable potential yields to be achieved. Farmers around the world are aware that farm-level land and water management practices are of prime importance for satisfying the needs of field-crop and other agricultural and horticultural ecosystems. Therefore, they endeavour to optimise the water supply of their crops within the limits of their knowledge and the farming operations practised. That is, over time, they have developed some sort of on-farm water management (OFWM) practices. However, farmers may often be unaware that conditions for the operation of farms are changing continuously. This has consequences for farm-level water management as the improvement of OFWM is a never-ending process.

Water management can be defined as the planned development, distribution and use of water resources in accordance with predetermined objectives while respecting both the quantity and quality of the water resources. It is the specific control of all human interventions concerning surface and subterranean water. Every planning activity relating to water can be considered as water management in the broadest sense of the term. Therefore, OFWM can be defined as the manipulation of water within the borders of an individual farm, a farming plot or field. For example, in canal irrigation systems, OFWM starts at the farm gate and ends at the disposal point of the drainage water to a public watercourse, open drain or sink.

OFWM generally seeks to optimise soil-water-plant relationships in order to achieve a yield of desired products. The managers usually try to achieve this desired yield by minimising inputs and maximising outputs, so as to optimise profits. In order to accomplish this, water has to be managed skilfully through certain practices covering areas of: soil and water conservation, water application, drainage, soil amelioration, and agronomy. All this has to be done within the context of the socio-economic environment of the community and the farmer's personal situation. There are a range of tools available that enable the manager to apply these practices.

When defining OFWM, it becomes clear that the term covers not the water resources, the irrigation facilities, the laws, the farmers' institutions, the procedures and the soil and cropping systems. OFWM is concerned with how these tools and resources are used and made available to provide water for plant growth. Moreover, it encompasses all the water used for that purpose, i.e. precipitation and water applied through irriga tion. Furthermore, it includes the use of respective practices and tools to improve site conditions and to protect crops and farming property from excessive water.

In the past, extensive research work in many countries has shown that, for example, in irrigated agriculture, good OFWM practices require well-levelled fields, appropriately designed on-farm distribution systems, and a good knowledge of when to irrigate and how much water to apply. Irrigated agriculture also requires a reliable source of water, readily available when needed, and in quantities that can be distributed effectively and efficiently over the farmer's field.

In addition, soil amelioration measures such as sub-soiling and gypsum application may be necessary, as may a well-functioning drainage system. However, all this will lead to good OFWM only if the system as a whole is well managed, if the managers or farmers take the appropriate management decision at the right time, and if they make sure that the management decisions are indeed transferred into practice correctly and on time. Furthermore, part of the success of good OFWM depends on close communication and interaction among farmers and other water users of the respective catchment area as well as with the service providers and the water supply administration. This is especially the case if the measures taken on-site will be affected by off-site activities or will affect such activities.

To satisfy the objectives of water control, an adequate soil water content must be maintained in a field. This must be done in a timely manner and within constraints imposed by characteristics of the area.

— Physical —soil types, depths, and characteristics, field layout, water sources and sinks.

— Climatic—frost potential, drought potential, rainfall amounts and intensities.
— Economic—market prices, material availability, labor cost and availability.
— Social—governmental regulations, environmental concerns, safety considerations.

Water budgeting and water management evaluations are useful tools to help the manager operate the system effectively and efficiently. Water budgeting is an accounting for all water entering, leaving, and remaining on a farm or field. This includes knowing the designated use of the available water as well as the losses from the system that cannot be controlled by the manager. Water management evaluations examine the efficiency with which water is conveyed, applied to a field, and made available for crop use for optimum growth. A well-managed system will, under most circumstances, not subject a crop to either drought or flood conditions that would result in yield reductions.

Components of Water Management System

The term "management" implies that the on-farm water management concept is one involving various decision-making processes. These decisions revolve around using water to realise optimum returns from the crop. Economic and social factors place certain restrictions on optimum returns. Components of the water management system that must be managed on a farm can be broken down into several categories.

— Irrigation
— Drainage
— Conveyance to and from fields
— Water storage
— Release of water from the confines of the farm
— Water sources and sinks

Relationships between the components are shown schematically in Figure 1. To satisfy the overall objectives of the water management program, decisions must be made concerning both the types of system components to instar, as well as how, when, and how long each system should be operated. These decisions must conform to both government and other social regulations on volumes of water entering and leaving a farm.

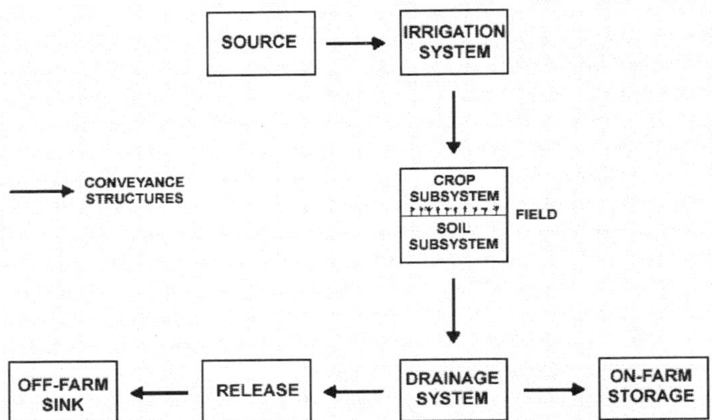

Figure 1. Relationship between systems components

Irrigation

The primary purpose of irrigation is to supply crops with sufficient water to satisfy evapotranspiration needs so yield reductions will not occur. The first management decision regards the type of system that will be used to perform this function. Common irrigation systems used are varied.

— Seepage or subirrigation

— Sprinklers — big gun, center pivot, linear move, or solid set

— Surface — furrow or crown flooding

— Trickle or low volume-subsurface, spray, drip emitter, bubbler, or mist

When to irrigate and how much water to apply are other concerns. Both questions can be answered by considering the evapotranspiration needs of the crop, rainfall patterns, and the need for field access. Clearly, water must be applied to the root zone before drought stress can cause irreversible damage to the crop.

When rainfall occurs in adequate amounts during major portions of the crop season, irrigation is used on a supplemental basis (in the event that rainfall is not timely or adequate). In this case, the cost of installing a sophisticated irrigation system must be weighed against the level of protection that the irrigator wishes to maintain. Irrigations must also be scheduled around the times needed to gain access to fields in order to perform cultural practice activities.

Other reasons for irrigation may also exist. Three such reasons are frost protection, pest management, and salinity control. Additional considerations are the coexistence of the irrigation system with cultural practices necessary for crop growth, the permanence of the system, and the source of water. The latter item is extremely important because quality and quantity of water may preclude the use of a particular irrigation method.

Drainage

In areas of high rainfall, adequate drainage is mandatory for crop production. In more and more regions, drainage may still be necessary to remove any excess irrigation water applied. The function of a drainage system is to remove water from a field root zone whenever necessary, and fast enough to avoid permanent root damage resulting from lack of oxygen in the soil.

Drainage systems must be compatible with crops grown, field layouts, and cultural practices such as crop rotation and cultivation. System choices include open ditches, tile drains, mole drains, and land forming for increased surface runoff. The four methods can be used separately or in different combinations. The system design must be capable of removing excess water at a high enough rate to ensure a level of protection against flood damage to crops.

Drainage system choices and operational decisions are limited by timely access to the field, economically advantageous field sizes, depths of root zones, depths of soil profiles, natural water table levels, amounts of water to be removed in a given time frame, and susceptibility of the particular crop to flooded conditions. Since drainage structures are relatively permanent, the system installed should be capable of handling the needs of all crops included in the grower's rotation.

Conveyance

Conveyance structures transport water to and from various locations on the farm, and include laterals, manifolds, submains, and mains. These delivery structures can be either open ditches or closed conduits, depending on the water source and the irrigation and drainage systems. The primary purposes of these structures are to transport water from the source (either surface or ground water) to the irrigation system, or from the field to the sink where runoff is deposited. A typical field conveyance system layout is shown in Figure 2 . A particular field does not need every component of the conveyance system. Devices such as furrows or other open ditches, sprinkler heads and drip emitters draw water from the laterals and spread it over the field surface.

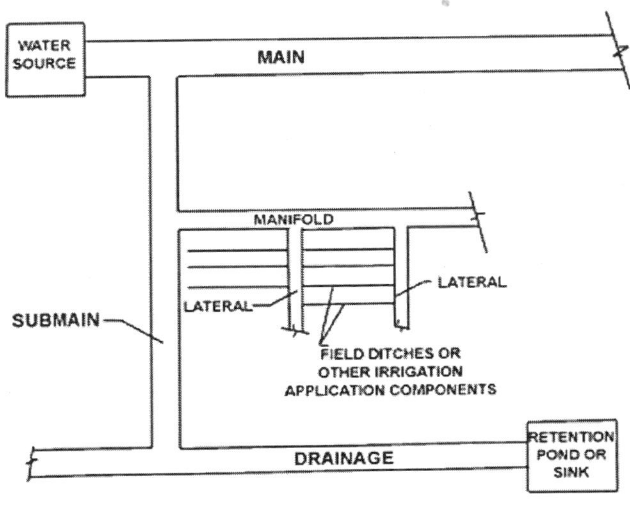

Figure 2. Typical field conveyance system layout

The conveyance structures must be large enough to handle the maximum irrigation or drainage volumes foreseen for the areas they serve, given a certain desired level of protection from flood and drought. Pressurised irrigation systems will generally require closed conduits for at least part of the conveyance or delivery system. Surface systems may use buried closed conduits to transport water in order to avoid taking land out of production. For drainage systems, open ditches, closed conduits attached to subsurface drain lines, or a combination of both may be used. If a combination is used, open ditches could carry surface runoff and subsurface drain discharge to the appropriate on- or off-farm sink.

The selection of a conveyance system should be based on land availability, access to parts of the farm (open ditches may impede ready access), volumes of water handled (burying very large conduits may be infeasible as well as expensive), rainfall amounts and intensities, water source, and type of irrigation system used. The manager or installer must then decide on how, when, and how much water will need to be routed to different parts of the farm to determine what the capacities of the structures should be. These decisions must be made to ensure that the crop is irrigated and drained in a timely fashion. To do this, a manager must know crop water needs, areas served by each structure, and the time that it takes to transport a certain volume of water in the system from one point to another.

Storage

Water can be stored on farms in various ways. The conveyance system generally provides a certain amount of storage to hold drainage amounts in excess of that removal amount planned for during design. Holding excess water in the conveyance structure may result in inadequate hydraulic head differences for drainage, or excessive seepage into a field.

The soil root zone is also a viable storage facility in which water in excess of crop needs can be held for short periods of time. The amount of water that a soil can store will depend on the soil type and the flood tolerance of the crop, both of which provide only a limited amount of storage for. a limited amount of time.

On a larger, long-term basis, on-farm reservoirs or ponds can provide water for irrigation either as primary sources or to augment the primary source in times of need. These reservoirs or ponds can hold agricultural runoff for later use, thus reducing the demand on the basin-wide system in times of drought. These ponds hold nutrient-laden water for reuse and also prevent degradation of public water supplies.

The selection of an on-farm storage facility is not always necessary. However, storage facilities provide the grower with additional insurance in the event of unduly large rainfalls or drought conditions. The use of these reservoirs is also restricted by the topography of the land and the type of irrigation system used.

For example, in areas where shallow soils overlying bedrock exist, or in areas where there are no natural catchment areas, the construction of a reservoir would be costly and only marginally beneficial since large amounts of land would have to be taken out of production. This is especially true in cases where irrigation systems require the movement of relatively large volumes of water through leaky open ditches. However, the reservoir is the most useful and powerful on-farm storage system under suitable conditions. Soils and conveyance structures can provide only temporary, small-scale relief for both irrigation and drainage needs.

Release

The final component of the on-farm water management system is the manner in which water is released from the confines of the farm. Some water will seep out of the system through the soil or percolate beneath the root zone. Water will also exit the system through evaporation from open conveyance or storage structures. However, to remove large volumes of water from a farm requires a structure capable of artificially channeling water to a point where it can either flow by gravity or by pumping to a sink outside the system. This is particularly important in areas of high rainfall.

Pumps can be used to move large amounts of water quickly. Alternatively, ditches or conduits on grades, and positioned above the water level of the sink, can perform the function adequately. Again, these structures must have the capacity to remove enough water, fast enough, to prevent crop damage given a selected level of protection.

The water manager must then make decisions for releasing water based on the drainage system capacity, crop tolerance to flood or drought, level of insurance desired in the event of needing more or less water on the farm in the future, and governmental or other special restrictions placed on water release to the public water supplies. Regulations and restrictions often involve limitations on the quality of water released to the public water supplies. The amount of nutrient-laden water that leaves a farm should be of concern to the grower, since it may be possible to reuse that water to his benefit.

Source and Sink

Two additional components of a farm water system are the water source and the sink. The source can be surface water from reservoirs, lakes, ponds, and canals, or ground water from wells. The sink can be on- or off-farm ponds, reservoirs, lakes, or canals. Once properly cleansed, excess water may also be used to recharge deep aquifers. Either gravity flow or mechanical pumping can be used to obtain water from the source and to pass runoff to the sink. In both cases, water quality must be considered an extremely important consideration.

Problems in On-Farm Water Management

In addition to satisfying the needs of field-crop ecosystems, OFWM is of prime importance to soil and water conservation. However, field observations have shown that not all forms of OFWM are appropriate to achieving a sustainable land use while conserving soil and water. For example, differences in erosion due to different management practices of the same soil are often greater than the differences in erosion from different soils under the same management.

The same applies to water where differences in water use efficiency due to different management at the same site are greater than the differences in water use efficiency at different sites under the same management. Although inappropriate management practices are an old problem in OFWM, this problem now ranks higher on the agenda as the world is faced with significant population increases, as food security becomes more of an issue and as land and water resources become scarcer.

Traditional farming systems in the tropics and subtropics often included the 'good practices' of water and soil conservation. However, in many cases, increased population pressure, the introduction of new cash crops and farming systems, and the mechanisation of farming operations have led to the abandonment of these farming and conservation practices. The shift towards annual crops on steep slopes that were previously under forest, tree crops or permanent pasture has led to increased water and soil losses. Other practices such as ridge constructions, ploughing down slopes and clean weeding have created additional problems in this respect.

Problems associated with the introduction of irrigation are: waterlogging; salinisation; soil and water contamination; and a lowering of the quality of the water released through natural or artificial drainage. Overirrigation is a common fault that together with seepage from unlined canals causes the groundwater table to rise. This leads to waterlogging and salinisation especially where there is no natural drainage or where the drainage system is inadequate. In addition, yields of crops commonly grown at the site are also reduced if the soil is compacted and soil aeration and water permeability are reduced.

Another old problem is the overuse of groundwater. The new aspect of this problem is that it has become increasingly widespread. Overexploitation leads to lower ground-water levels and increasing costs of supply, and, under certain circumstances also to subsidence, landslips, seawater intrusion, etc. All these problems have been well known to agriculture for some time. They are mostly the result of inappropriate land use decisions and/or the inability to adjust to changing circumstances. They are quite often influenced significantly by non-agricultural factors such as those of a social, economic or political nature. In many cases, political factors are more important than the technical considerations usually discussed in connection with onfarm land and water management issues.

Traditional management practices of the irrigation supply and conveyance systems often contribute to high water losses. On many farms, the low irrigation efficiency is further accentuated by farmers' traditional irrigation methods and practices, inadequate land levelling, lack of a crop-specific water application, insufficient drainage, and poor maintenance of irrigation and drainage infrastructure.

Farmers are often unaware of the possibilities of applying water in a more productive way. The potential of horticultural crops with their high land, water and labour productivity is often not adequately recognised, especially by less educated and poorer farmers. Farmers generally lack technical and economic information on improved OFWM methods and techniques and on the related aspects of more productive cropping patterns and crop management. Therefore, proper training and capacity building at all levels of OFWM would be useful.

As mentioned above, these problems have been a threat to agriculture in many countries in past decades. However, these problems have now taken on an added dimension. There is increasing public demand that the development and management of land and water be ecologically sustainable as well as economically viable. As there is a close interrelationship between land use and water resources, farmers have to be aware of this interrelationship and have to adjust their OFWM efforts to address this issue.

Hence, both on-site and off-site effects have to be considered in management planning and practices. By increasing the proportion of rainfall lost due to surface runoffas a result of inappropriate land use, problems of flooding and downstream erosion arise following rain events. As less water percolates down to provide base flow for streams, the dry season flow may be lowered and even reduced to zero. Lowering of the water table can lead to the loss of well water. Decreased soil-moisture supplies result in progressively poorer vegetation. Removal of the fertile topsoil by erosion will reduce crop yields.

Deposition of coarse sand, gravel and stones removed from steep slopes onto low-lying areas decreases the agricultural potential of the soils.

Sediment deposition in channels and reservoirs resulting from upstream erosion is causing major problems. Suspended sediment represents a deterioration in water quality. Fine clay particles require expensive treatment by chemical flocculation and filtration. The contamination of water resources by agriculture with harmful substances has become a widespread problem and one of increasing concern for the nonagricultural public.

Another new feature is the fact that the agriculture sector is coming under pressure to make more efficient use of water. It has been and still is criticised for being the greatest water user while having the lowest water use efficiency and lowest output per unit of water used of all sectors. In particular, irrigated agriculture, the greatest water user of all, has been accused of being responsible for inefficient water use and land degradation. In the past, agricultural research and practices dealt solely with the subject of water for the purpose of optimising water management in order to satisfy the water needs of the crop.

It is only in recent times that the effects of agricultural activities on water resources have been studied in the context of intersectoral water management and have consequently become a matter of public concern. In areas of water scarcity, the competition between different sectors of water users is attracting increasing public attention. It is also clear that future agricultural practice and research will no longer be geared exclusively to the task of optimising the water supply of crops. Rather, there is a need to curb agricultural water consumption for the benefit of other sectors, for example, by reallocating water originally designated for crop production to supplying drinking-water and domestic water to the population. The challenge is clear: OFWM has to contribute to an increase in overall water use efficiency.

Finally, it is necessary to realise that the water issues of quantity and quality are not related solely to agricultural production. They are increasingly related to urbanisation and industrialisation. The migration of water from agriculture to urban and industrial uses is underway and increasing, driven by the fact that "water flows uphill towards money". Aggregate supply economics suggest that cities will gain in the long run because of the higher prices for water that urban users are more likely to pay.

Within the next few decades, agriculture will face a number of important challenges. On the one hand, it is necessary to increase food and fibre production in order to guarantee food and clothing for the increasing world population. This contribution is urgently needed to help combat poverty while fostering economic development. On the other hand, agriculture will face increasing competition for decreasing water resources.

The situation is exacerbated by the dwindling financial resources and increasing costs for the rehabilitation of existing irrigation and drainage systems and the setting up of new ones. Moreover, agriculture needs to meet these challenges in a political and social environment that, in many countries, is highly critical of agriculture in general and of irrigated agriculture in particular. Agriculture needs to be more sparing in its use of water resources, minimise its impact on the environment and exercise continual self-restraint by means of eco-auditing.

Until now, it has been farmers who have decided what happens on agricultural land. They make rational decisions according to their own circumstances. Their decisions are influenced by: physical factors, such as soil and climate; the socio-economic features of the community; and their own personal situation. Technical advice and the assistance available may be another influencing factor. As natural resources grow scarcer, the general public sees the environment as being increasingly endangered. Hence, farmers will face the challenge of land use decisions being influenced increasingly by non-agricultural factors with pressures coming from social, economic and political quarters.

Farmers cannot ignore the fact that agricultural activities produce a significant proportion of all pollutants entering streams, lakes, estuaries and groundwater. This is especially the case in Europe, North America and Australia, but it is also increasingly true in developing countries.

Farmers need to admit that, in order to solve the water pollution problem, agricultural non-point source (NPS) pollution has to be controlled. They have to be aware that the solutions to controlling runoff will require an integrated effort by landowners, government and organisations responsible for protecting and restoring soil and water. Osmond et al. see the first step in reducing agricultural NPS pollution as being one of focusing on the primary water-quality problem within the watershed; the water-quality use impairment must be identified and the type and source of the pollutant(s) defined. Once the problem has been defined clearly and documented, the critical area, i.e. the area that contributes the majority of the pollutant to the water resource, can be identified. Land treatment, that is the installation or utilisation of good management practices (GMPs) or better GMP systems, should then be implemented on these critical areas. Good and effective OFWM adjusted to the individual circumstances is an essential part of the GMP systems. For example, in some cases it might be important for farm runoff to be kept on the farm area where there is a risk of nutrient or pesticide contamination. This will require special OFWM efforts.

In many cases, the potential for achieving benefits from improved OFWM is substantial. Water savings, yield increases, higher productivity and higher farmer incomes are achievable. Many experiments have demonstrated a positive effect of the yields for farmers. However, in many cases, questions concerning the ability and willingness of farmers to adopt the improved practices remain. In this context, the lack of financial and economic impact assessments of various interventions and the effects of improvement measures on productivity, incomes and water use efficiency remains a major shortcoming, also for policy and strategy formulation.

Water as a means for production will become increasingly scarce and expensive, making high on-farm water use efficiencies and precise water management indispensable. Furthermore, farmers may be forced to devote more attention to the sustainability of natural resources on their land and within their catchment area. This may require and prompt changes in OFWM practices. In this respect, the challenge for farmers will be to manage the water on their farms professionally in the most sustainable and profitable way. This will require detailed planning and a clear definition of the goals and the means for achieving them within the constraints given. A sound understanding of the environment and a clear commitment to sustainable OFWM practices are essential. In order to achieve all the above, it will be necessary to implement good information management and flows. In this way, it will be possible to supply the necessary information and data on which decisions should be based.

For accomplishing GMPs in OFWM, the management structure of the individual farm is quite important. This is especially the case where the labour process is becoming fragmented in the course of development. For example, in Egypt, the head of a rural household is primarily a manager, linked to other households through exchanges and the hiring of labour, rental of machinery and land, and other relationships. One of the tasks of the head of the household is to manage the labour input in agriculture. The head of the household is also usually the only household member who follows the crop through the crop cycle. The other household members and the hired labour do most of the physical work and often lack comprehensive knowledge of crop and water management. The separation of management and farm labour causes a concentration of knowledge in the person of the manager. Hence, a de-skilling of the farm labour takes place, a fact which needs to be considered when developing and implementing strategies for the introduction of advanced OFWM practices.

With generally rising income expectations and standards of living, increased agricultural yields become necessary to satisfy the higher income demand. This is why, in the course of development, land which has been regarded as fairly fertile up to this point will become marginal land, and the previously marginal land will go out of production as it may no longer guarantee the necessary yield and income levels. Agriculture and especially irrigated agriculture will concentrate increasingly on highly productive land only. More food will be produced on an even smaller area. Achieving this requires more sophisticated management of the production technologies in general, and more advanced OFWM in particular.

The implementation of a more advanced form of OFWM constitutes a major challenge for the extension service. The conventional approach of the existing extension service has often proved not to be very successful in changing farmer behaviour. This has mainly been because extension workers often have little awareness about farmers' actual needs and problems, or about the practical value of their messages with regard to the farmers' social and financial conditions. In addition, they often lack experience. New and more participatory ways for the dissemination of information in the field of OFWM seem to be needed urgently.

A participatory group extension approach is currently being introduced in Egypt's extension system. It is based on the principle of learning and doing together. It seems to be the only alternative problemsolving concept conceivable for introducing a sustainable, advanced form of OFWM. Generally, it is necessary to conduct extensive education programmes in order to update continuously the knowledge base of the extension service staff and to encourage farmers to adopt good OFWM practices.

The introduction of an advanced form of OFWM also poses many challenges for the agricultural research community. For example, as not all forms of OFWM are good, there is a strong interest in establishing criteria of what constitutes good and bad OFWM. In this sense, OFWM investigations/research/projects cannot be primarily descriptive, although detailed descriptions of the management processes may be important. Instead, OFWM investigations/research/projects have to be predictive. However, the interest lies not only in predicting what a given management will do in a certain circumstance, but more fundamentally in predicting what would occur if certain activities were adopted. The outcomes of these predictions need to be evaluated and an attempt made to rank management activities in terms of better or worse and to discern the best if possible.

Agricultural Water Reuse

In areas where irrigation water is scarce, the use of drainage water is an important strategy for supplementing water resources. The reuse may help alleviate drainage disposal problems by reducing the volume of drainage water involved. The reuse of drainage water for irrigation can reduce the overall problems of water pollution. Reuse measures consist of: reuse in conventional agriculture; reuse to grow salt tolerant crops; reuse in wildlife habitats and wetlands; and reuse for initial reclamation of salt-affected lands.

Drainage water is normally of inferior quality compared to the original irrigation water. Adequate attention needs to be paid to management measures to minimise long-term and short-term harmful effects on crop production, soil productivity and water quality at project or basin scale. The drainage water quality determines which crops can be irrigated. Highly saline drainage water cannot be used to irrigate salt sensitive crops, but it can be used on salt tolerant crops, trees, bushes and fodder crops. A major concern in reuse measures is that drainage water from reused waters is often highly concentrated, requiring careful management.

Drainage water quality is the major concern in reuse possibilities as it defines which crops can be irrigated and whether long-term degradation of soil productivity is a major issue. On the other hand, the soil type, drainage conditions of the land and the crop salt tolerance define what quality drainage water can be used for irrigation in combination with the availability of other freshwater resources.

Pollutants from surface runoff, i.e. sediments, pesticides and nutrients, play a minor role in reuse for crop production. However, for sustainable agricultural practices and to prevent environmental degradation, nutrients supplied with reused drainage water should be deducted from the fertilizer requirements in order to prevent imbalanced and excessive fertilizer application.

Subsurface drainage water generally shows increased concentrations of salts and sometimes certain trace elements and soluble nutrients. Salts and trace elements play a major role in the reuse of drainage water. Above a certain threshold value, high total concentrations of salts are harmful to crop growth, while individual salts can disturb nutrient uptake or be toxic to plants. A high sodium to calcium plus magnesium concentration ratio may cause unstable soil structure.

Soils with unstable structure are subject to crusting and compaction, degrading soil conditions for optimal crop growth. Toxic trace elements such as boron can interfere with optimal crop growth and others such as selenium and arsenic can enter the food chain when crops are irrigated with water containing high concentrations of these trace elements. This is of major concern for human and animal health.

Municipalities and industries often use agricultural main drains to dispose of their wastewater. In many areas around the world, municipal and industrial wastewater is either insufficiently treated, or not treated at all. Bacteriological and organic and inorganic compounds seriously pollute main drains, posing an environmental hazard to both human and wildlife and restricting reuse from main drains. For example, in the Nile Delta, Egypt, the discharge of untreated wastewater is a major concern for the sustainability of an irrigated agriculture that depends heavily on the large-scale reuse of agricultural drainage water to meet water shortfalls.

The biochemical oxygen demand (BOD) is defined as the amount of oxygen consumed by microbes in decomposing carbonaceous organic matter. The chemical oxygen demand (COD) is the amount of oxygen required to oxidize the organic matter and other reduced compounds. The high chemical versus biological oxygen demand (COD/BOD) ratios imply significant industrial pollution. Total coliform is the most probable number of faecal coliform (MPN) in 100 ml and a high value indicates severe pollution from municipal sewage water.

The degraded water quality threatens the expansion and even the continuation of the reuse of drainage water from the main drains (official reuse) in the Nile Delta. Since the 1990s, many reuse mixing stations have been under increasing pressure of water quality deterioration. Indeed, since 1992, 7 of the 23 main reuse mixing stations have been entirely or periodically closed.

Due to the increasing deterioration of water quality, new opportunities for reuse of drainage water are being explored. Between the centralised official reuse and the localised unofficial reuse (where farmers pump drainage water from the collector drains directly onto fields), there is the option to capture drainage water from branch drains and pump it into the branch canals at their intersections. This level of reuse, termed intermediate reuse, captures relatively good quality drainage water before it is discharged into the main drains where it is lost for irrigation because of pollution from other sectors. This problem highlights the need for catchment level planning to protect and use all water resources sustainably.

Maintaining a Favourable Salt Balance

The major concern in the reuse of agricultural drainage water is the buildup of salts and other trace elements in the rootzone to such an extent that it interferes with optimal crop growth and degradation of the aquifers. Applying more water than necessary during the growing season for evapotranspiration can leach the salts. In areas with insufficient natural drainage, leaching water will need to be removed through artificial drainage. As safe disposal of agricultural drainage water is often the hindrance to sustainable drainage water management, the solution in reuse lies in applying just enough water to maintain a favourable salt balance.

Maintaining a Favourable Soil Structure

The sodium hazards of irrigation water are related to the ability of excessive sodium or extremely low salinity concentrations to destabilise soil structure. The primary processes responsible for soil degradation are swelling and clay dispersion. Provided that the salt concentration in the soil water is below a critical flocculation concentration, clays will disperse spontaneously at high exchangeable sodium percentage (ESP) values, whereas at low ESP levels inputs of energy are required for dispersion.

The salt concentration in the soil water is crucial to determining soil physical behaviour because of its effects in promoting clay flocculation. However, the boundary (ESP/salt concentration of the soil water) between stable and unstable conditions varies from one soil to the next and changes with the clay mineralogy, pH, soil texture, and clay, organic matter and oxide content. The short-term effects of irrigating with water having excess sodium or very low salt concentrations relate mainly to infiltration problems. The reduction of infiltration can be attributed to the dispersion and migration of clay minerals into soil pores, the swelling of expandable clays and crust formation. The potential infiltration problems in relation to the quality of the irrigation water are normally evaluated on the basis of the salinity and SAR of the irrigation water.

Various authors have developed stability lines related to the total salinity concentration and the SAR. The actual line that represents the division between stable and unstable soil conditions is unique for each soil type and varies with soil conditions. Published guidelines on infiltration problems in relation to the SAR and the salinity of the applied water can therefore provide only approximate guidance. Where large-scale reuse of drainage water is planned and sodium hazards might be expected, stability lines need to be established for local conditions.

Infiltration problems caused by the sodicity of irrigation water also depend on irrigation and soil management. Especially at lower SAR levels when chemical bonding is weakened, but no spontaneous dispersion takes place, inputs of energy are required for actual dispersion. For example, sprinkler irrigation increases the likelihood of surface crusting due to the high physical disruption as the drops hit the soil surface aggregates. In soil management, incorporation of organic matter increases the stability of the soil aggregates and reduces the hazards of structural degradation as a result of sodicity.

Reduction of the hydraulic conductivity in the soil profile is normally a long-term effect resulting from the use of sodic water. In particular, the presence of carbonates and bicarbonates in the water could result in soil degradation in the long term because precipitation of calcium carbonate increases soil SAR. Calcium in the form of calcite is one of the first salts to precipitate. Upon further concentration magnesium salts will also precipitate.

There is little need to undertake action to increase infiltration and or hydraulic conductivity unless crop water or leaching requirements cannot be met or if secondary problems reduce crop yields or impede seedling emergence. Secondary problems include crusting of seed beds, excessive weed growth and surface water ponding that can cause root rot, diseases, nutritional disorders, poor aeration and poor germination.

Management options to mitigate and reduce these problems can be chemical, biological and physical. Chemical management options entail adding chemical amendments to soil or water and thereby changing the soil or water chemistry. The aim of biological methods is to improve soil structure or to influence the soil chemistry through the decomposition of organic materials. Physical methods include cultural practices to increase infiltration rates during irrigation and rainfall or to prevent direct contact between ponding water and plant stems, roots and seeds.

Chemical Soil and Water Amendments

The aim of applying chemical amendments to soil or water is to improve poor infiltration caused by either a low salinity or by excessive sodium. The problem is most severe at low electrolyte concentration and high SAR. Improvements can be expected if the soluble calcium content is increased or a significant increase in salinity is achieved.

Most soil and water amendments supply calcium directly or indirectly through acid that reacts with soil calcium carbonates. Acid is not effective where calcium carbonate is not present in the soil profile. However, calcium carbonate is often present in arid soils. Water amendments are most effective when infiltration problems are caused by low to moderate saline water (EC < 0.5 dS/m) with a high SAR. Where moderate to high saline water (EC $>$ 1.0 dS/m) causes problems, soil amendments are more effective.

Amendments need to react quickly if they are to solve actively infiltration problems caused by irrigation water. Many of the amendments for improving or reclaiming sodic soils react only after oxidation. Thus, they are slow and less suitable for solving infiltration problems caused by irrigation water. Slow reacting amendments include acid forming substances such as sulphur, pyrite, certain fertilizers and pressmud from sugar-cane factories.

Gypsum ($CaSO_4.2H_2O$) is the most commonly used amendment. It can either be applied to the soil or irrigation water. Infiltration problems normally occur primarily in the upper few centimetres of the soil. Hence, it is normally more effective to apply small frequent doses of gypsum left on the soil surface or mixed with the upper few centimetres of the topsoil than higher doses incorporated deeper in the soil profile.

The application of gypsum to irrigation water to prevent infiltration problems usually requires less gypsum per hectare than when it is used as soil application. Gypsum application to water is particularly effective when added to low salinity water (< 0.5 dS/m) and less effective for higher salinity water because of the difficulty of obtaining sufficient calcium in solution. In practice, it is not possible to obtain more than 1-4 meq/litre of dissolved calcium, or 0.86-3.44 g/litre of pure gypsum in fast-flowing irrigation streams. In low salinity water, these small amounts of dissolved calcium ions may increase the infiltration rate by as much as 300 percent. To use gypsum as a water amendment, finely ground gypsum (< 0.25 mm in diameter) is preferred as it has a higher solubility. This is also normally the purer grade of gypsum. A drawback is that finely ground purer grades of gypsum are more expensive, which often prevents small farmers from using it. The coarser grinds and lower grades of gypsum are more satisfactory for soil application.

Experiments have been conducted with the placement of gypsum rocks in the watercourse. The problem is that the amount of calcium dissolving from the gypsum rocks is low so the effectiveness depends on the stream velocity and volume. Experiments from Pakistan have shown that sufficient head and length of the supply canal from the tubewell to the watercourse are required to dissolve enough calcium in tubewell water. Where technically feasible, the use of gypsum stones has proved to be financially attractive for farmers. In large-scale applications, a drawback might be the cost of canal maintenance, as the gypsum blocks need to be removed before mechanical cleaning.

Sulphuric acid is an extensively used amendment for addressing infiltration problems. It is effective only where lime is present in the soil surface. This highly corrosive acid can be applied to the soil directly, where it reacts with lime making the naturally present calcium available for exchange with adsorbed sodium. Sulphuric acid reacts rapidly with soil lime making it a useful amendment to combat infiltration problems. When added to the irrigation water, it is neutralised by the carbonates and bicarbonates in the water and any excess would contribute towards dissolving the soil lime.

Biological Soil Amendments

Crop residues or other organic matter left in or added to the field improve water penetration. The more fibrous and less easily decomposable crop residues are more suitable for mitigating minor infiltration problems. Fibrous organic materials keep the soil porous by maintaining open voids and channels. The use of crop residues forming polysaccharides as a cementing agent is most effective where they are incorporated only in the upper few centimetres.

Incorporation of easily decomposable organic matter and incorporation deeper into the soil profile do not generally help reduce infiltration problems although they do improve soil structure by producing polysaccharides that promote soil aggregation and enhance soil fertility. Another important aspect of the decomposition of organic matter is the production of carbon dioxide, which in turn increases the solubility of lime.

Cultivation Practices

Cultivation is usually done for weed control or soil aeration purposes rather than to improve infiltration. However, where infiltration problems are severe, cultivation or tillage are helpful as they roughen the soil surface, which slows down the flow of water, so increasing the time during which the water can infiltrate. Cultivation is only a temporary solution. After one or two irrigations, another cultivation may be needed. Moreover, the construction of (broad) beds may help mitigate the ill effects of standing water as it prevents direct contact of the plants with the water. Research from Pakistan has shown that cotton in particular benefits from planting on broad beds.

Maintaining Favourable Levels of Ions and Trace Elements

High concentrations of trace elements in soil, ground and drainage water can occur in association with high salinity and can be affected by the same processes. However, in some places they may also occur independently of salinity. In examining ways to control levels of ions and trace elements in the rootzone, it is necessary to understand the processes that affect their mobility.

Deverel and Fujii provide a framework for evaluating concentrations of trace elements in soil and shallow groundwater. The two processes that largely control the mobility of trace elements in the soil water are:

i) adsorption and desorption reactions; and

ii) solid-phase precipitation and dissolution processes.

These processes are influenced by changes in pH, redox state and reactions, chemical composition and solid-phase structural changes at the atomic level.

There are no well-tested and simple models for estimating changes in trace element concentration as a result of irrigation and drainage water management. Nor have irrigation water quality criteria for trace elements been established. However, guidelines have been developed for trace elements based on results from sand, solution and pot trials, field trials with chemicals and laboratory studies of chemical reactions.

Selenium

Albasel et al. reviewed the quality criteria for trace elements in irrigation water compiled by the National Academy of Engineering in 1973. The recommendation for selenium was that the concentrations in irrigation water should not exceed 20 mg/litre. The guideline was recommended for all irrigation water on any land without consideration for soil texture, pH, plant species, climate and other water characteristics such as sulphate concentration. In the San Joaquin Valley, specific conditions indicated that a review of this guideline was required:

i) selenium in the drainage water is in the form of selenate, which is not readily adsorbed onto soil particles and is thus readily leached;
ii) water containing high selenium concentrations also has a high total salinity content and thus requires high leaching fractions to prevent salinity build up in the soil profile; and
iii) water containing selenium has high concentrations of SO_4^- that greatly inhibit plant uptake of selenium. Albasel et al. used the concentration fractions for various LFs to convert the concentration of selenium in the irrigation water (Se_I) into selenium concentration in the soil solution (Se_{ss}).

Research with alfalfa, which is the most sensitive crop to selenium accumulation in relation to the use of the harvested product, showed that Se_{ssm} is 250 mg/litre without exceeding the Sehp of 4-5 mg/kg if the saline irrigation water is dominated by sulphate. Assuming that the *LF* would be 0.2 or more, the Se_I could be 100 mg/litre. Where other crops are grown on soils different from those in the San Joaquin Valley, new guidelines will be necessary. These should be based on soil and water properties and the maximum Se_{hp} specific for those crops and the use of the harvested product.

Boron

Boron is an essential micronutrient for plants but it is toxic at concentrations only slightly above deficiency. The range of boron tolerance varies widely among crop plants. Salt sensitive crops such as citrus, fruit and nut trees are sensitive to boron while salt tolerant crops such as cotton, sugar beet and Sudan grass tolerate higher levels of boron.

Leaching of soil boron is more difficult than soluble salts such as chloride. This is because of the slow dissolution of boron minerals and desorption of boron adsorbed to oxides of iron and aluminium in the soil. Hoffman established a relationship to calculate the relative decrease of soluble boron in soils during reclamation:

$$\left(\frac{B_{sst}}{B_{ss0}}\right)\left(\frac{D_L}{D_s}\right) = 0.6 \qquad (1)$$

where:

B_{sst} = desired boron concentration in the soil solution (mg/litre);

B_{ss0} = initial boron concentration in the soil solution (mg/litre);

DL = depth of leaching water (mm); and

Ds = depth of soil to be reclaimed (mm).

Keren et al. reported that native soil boron is more difficult to leach than boron accumulated from irrigation with boron-rich water. Where steady-state conditions exist between boron adsorbed and boron in the soil solution (B_{ss}), then the input and output of boron from the rootzone, and thus Bss, is related to the boron concentration in the irrigation water (*BI*) and the *LF*. To establish the relationship between *BI* and B_{ss}, a water uptake pattern of 40-30-20-10 percent from the first to the fourth quarter of the rootzone is assumed.

Water Reuse in Conventional Crop Production

Drainage water of sufficiently good quality might be used directly for crop production. Otherwise, drainage water can be reused in conjunction with freshwater resources. Conjunctive use involves blending drainage water with freshwater. Alternatively, drainage water can be used cyclically with freshwater being applied separately. In cyclic use, the two water sources can be rotated within the cropping season (intraseasonal cyclic use), or the two water resources can be used separately over the seasons for different crops (interseasonal cyclic use).

The choice of a certain reuse option depends largely on: drainage water quality; crop tolerance to salinity; and availability of freshwater resources. The quantity and time of availability of drainage water is of major importance. For example, where reuse takes place in an irrigation system in which water is distributed on a rotational basis, the probable mode of reuse is either direct or cyclic.

Direct Use

The direct use of drainage water is implemented mainly at farm level, whereby the drainage water is not mixed with freshwater resources. Research results from India, Pakistan, Central Asia and Egypt show that drainage water can be used directly for irrigation purposes without severe crop yield reductions where the salinity of the drainage water does not exceed the threshold salinity value for the crops grown and good drainage conditions exist. As crops are often more sensitive to salinity during the initial growth stages, research in India has revealed the importance of pre-irrigation with good quality irrigation water. Higher crop yields were attained when freshwater pre-irrigation was applied with only drainage water being applied thereafter. Under these conditions, drainage water with salinity levels exceeding the threshold value could be used whilst maintaining acceptable crop yields.

Conjunctive Use

Blending: Where drainage water salinity exceeds the threshold values for optimal crop production, it can be mixed with other water resources to create a mixture of acceptable quality for the prevailing cropping patterns.

Where reuse takes place by mixing drainage water from main drains with surface water in main irrigation canals, the most salt sensitive crop determines the final water quality. For example, according to the regulations for the Nile Delta, the maximum salinity of the blended water is rather low to ensure optimal production of the major crops grown, i.e. cotton, maize, wheat, rice and berseem. Maize and berseem are the most salt sensitive crops with an ECe threshold of 1.7 and 1.5 dS/m, respectively. To ensure potential maximum yield production of these two crops and assuming a concentration factor of two between the EC of the irrigation water and the ECe, the maximum allowable salinity of the irrigation water is about 0.8-0.9 dS/m.

Where mixing takes place at the farm level, the salinity of the blended water can be adjusted towards the salt tolerance of individual crops. Table 2 shows an experiment from India, where blended water with different levels of salinity was used to cultivate wheat.

Table 1. Drainage water quality criteria for irrigation purposes in the Nile Delta, Egypt

Salinity of drainage water (dS/m)	Restriction on use for irrigation
< 1.0	used directly for irrigation
1.0 - 2.3	Mixed with canal water at ratio 1:1
2.3 - 4.6	Mixed with canal water at ratio 1:2 or 1:3
> 4.6	Not used for irrigation

Table 2. Effect of diluted drainage water on wheat yield

ECI, dS/m	Relative Yield (%)	
	All irrigations	Post-plant irrigations
0.5 (canal water)	100	100
3.0	90.2	-
6.0	80.4	95.8
9.0	72.5	90.3
12.0	56.4	83.7
18.0	-	78.0

This experiment shows that for the case of all irrigations applied with a blended water mixture having a maximum salinity of 3 dS/m, a potential crop yield of 90 percent of the maximum yield was obtained. According to the crop tolerance data published by Maas and Grattan and assuming a concentration factor of 1.5 between water salinity and soil salinity, the maximum potential yield could be attained. The difference between theory and the actual field situation might be caused by a discrepancy in the actual concentration factor, which is the inverse of the LF. Moreover, the actual yields depend on many factors including physical, climate and farm management practices. In general, where all irrigations use the blended water, the maximum allowable salinity depends on the threshold value of the crops. Where the pre-irrigation uses freshwater, the blended water could have a salinity level exceeding the threshold value without affecting yields. The experiment from India shows that water used in post-plant irrigations with a salinity level of three times (ECI = 9 dS/m) the maximum allowable salinity level when only blended water is used (ECI = 3 dS/m) results in 90 percent yields.

The leaching of salts is necessary to maintain a favourable salinity balance in the rootzone. The procedures presented above might be used for this purpose. On the other hand, if higher salinity levels are tolerated towards the end of the growing period or if a more salt sensitive crop is grown after a more salt tolerant crop, the salinity levels have to be reduced sufficiently low in order not to interfere with the growth of the next crop.

Cyclic Use

Cyclic use, also known as sequential application or rotational mode, is a technique that facilitates the conjunctive use of freshwater and saline drainage effluent. In this mode, saline drainage water replaces canal water in a predetermined sequence or cycle. Cyclic use is an option for where the salinity of the drainage water exceeds the salinity threshold value of the desired crop. A condition for cyclic use is that two different water sources can be applied to the field separately. Therefore, it is not normally applied at irrigation-scheme level but at a tertiary or farm level. In India and Pakistan, the canal irrigation water is delivered on a rotational basis to the watercourses (tertiary canal) and individual farms. This offers considerable potential for cyclic use on tertiary or farm level. Modelling and field studies have demonstrated the feasibility of the cyclic reuse strategy.

The cyclic use of drainage water can be either intraseasonal or interseasonal. The latter mode of cyclic use follows the same principles for each cropping season as the direct use of drainage water.In intraseasonal cyclic use, the strategy implies that non-saline water is used for salt sensitive cropping stages and saline water when the salt tolerance of the plant increases. Such experiments have been carried out in India.

Saline drainage water (ECI of 10.5-15.0 dS/m) was combined with canal water for use in a pearl millet/sorghum-wheat rotation. In the experiment, canal water was used for pre-plant irrigations and thereafter four irrigations each of 50 mm depth were applied as per the planned modes to irrigate wheat. Pearl millet and sorghum were given pre-plant irrigation and thereafter did not receive further irrigation except by monsoon rains during the growth period. The results of the experiment support the cyclic use strategy.

No significant yield losses occurred in wheat when saline drainage water was substituted in alternate sequences (canal water - drainage water, or drainage water - canal water), or when the first two irrigations were with canal water and the remaining two irrigations were with drainage water, or when the first two irrigations were with saline drainage water and the remaining two irrigations were with canal water. Intraseasonal cyclic use offers considerable potential as not all crops tolerate salinity equally well at different stages of their growth. Most crops are sensitive to salinity during emergence and early development. Moreover, the flowering stage is also critical. Tolerance to salinity generally increases with the age of the crop. An exception is the salinity tolerance of mustard during the flowering to reproductive development stage.The long-term sustainability of this drainage water management option entails devoting sufficient to maintaining a favourable salt balance and to preventing a buildup of trace elements in the rootzone to levels toxic to plant growth.

Cyclic use also requires attention to soil degradation as a result of using sodic water. A high exchangeable sodium percentage on the soil exchange complex does not normally lead to soil degradation if it is compensated by a high soil moisture salinity to suppress the extent of the so-called diffuse double layer. However, upon irrigation with low saline irrigation water or rainfall, the diffuse double layer swells resulting in soil dispersion. Adding sufficient soil or preferably water amendments to compensate for the high sodium to calcium and magnesium content in the saline irrigation water can prevent these problems.

Crop Substitution and Reuse

Crops differ significantly in their tolerance to concentrations of soluble salts in the rootzone. The difference between the tolerance of the least and the most sensitive crops may be tenfold. A number of salt tolerant crop plants are available for greater use of saline drainage effluent. Raising the extent of the salinity limits through selecting more salt tolerant crops enables greater use of saline drainage effluent and reduces the need for leaching and drainage.

Reuse for Irrigation of Salt Tolerant Plants and Halophytes

Where the irrigation water is too saline to grow conventional agricultural crops, irrigation of halophytes might be considered. The maximum amount and kind of salt that salt tolerant plants and halophytes can tolerate vary among species and varieties. Halophytes have a special feature as their growth is improved at low to moderate salinity levels.

In contrast, salt tolerant crops have maximum growths up to a threshold salinity level after which growth is reduced. Salt tolerant plants and halophytes have been grown successfully in many places in the world to produce fuel, fodder and to a lesser extent food. Institutes in Australia have gathered useful salt tolerance data on a large number of trees and shrubs species. Growing salt tolerant plants and halophytes under saline conditions requires management of the salt balance in the rootzone. Where natural drainage is insufficient, artificial drainage is required to remove the leaching water.

Fuel: Many people in developing countries rely on wood for cooking and heating. As agricultural land is required to feed growing populations, it is unlikely that good quality agricultural land will be used for fuel production. Salt tolerant trees and shrubs can be grown for fuel production and building materials using saline water and marginal lands. Among the promising tree species for these purposes are Prosopis, Eucalyptus, Casuarina, Rizophora, Melaleunca, Tamarix and Acacia.

Fodder: Pasture improvement programmes in salt-affected regions throughout the world have used halophytes and salt tolerant shrubs and grass species. Trees and shrubs can be valuable complement to grasslands. They can serve as a nutrient pump and lower saline shallow groundwater tables. They are less susceptible to moisture deficits and temperature changes than grasses. They might also provide valuable complementary animal food or fuelwood. Salt tolerant grasses, shrubs and trees with potential for fodder use include:

Other products: Salt tolerant plants can also produce other economically important materials, e.g. essential oils, gums, oils, resins, pulp and fibre. Moreover, salt tolerant plants can be used for landscape and ornamental purposes and irrigated with saline water, thereby conserving freshwater for other purposes.

Reclamation of Salt-affected Land

Sodic soils often have low hydraulic conductivity as a result of the high sodium percentage on the soil exchange complex. The reclamation of sodic soils requires that a divalent solute (mainly calcium) pass through the soil profile, replacing exchangeable sodium and leaching the desorbed sodium ions from the rootzone. Therefore, the rate at which sodic soils can be reclaimed depends on the water flow through the soil and the calcium concentration of the soil solution.

The application of leaching water with a high electrolyte concentration promotes flocculation of the soils and thus improves soil permeability. This expedites the reclamation process. Amendments need to be added in order to replace sodium with calcium ions on the soil exchange complex. Over time, less-saline water needs to replace the saline leaching water in order to lower the salinity levels sufficiently to establish a crop.

The use of saline drainage water to reclaim salt-affected soils is not a permanent solution for reducing drainage effluent disposal volumes. It is only a substitute for the use of good quality irrigation water for reclamation purposes.

Drainage Water Disposal

Drainage water management is normally concerned with reducing the amount of drainage water and with managing its disposal. However, this aim is more complex than it appears. Drainage is practised to maintain aeration in waterlogged rootzone and/or to leach excess soil salinity to sustain agricultural production. The drainage water generated must then be managed for reuse purposes where it is of suitable quality and finally discharged or disposed of.

The discharge of drainage waters in watercourses may have impacts ranging from beneficial to deleterious. The disposal of drainage water into wetlands, lakes, rivers and coastal waters entails considerations about the quantity and quality allowable and indeed sometimes required to maintain desirable ecological conditions and functions of that given water body.

To sustain irrigated agriculture, the maintenance of favourable salinity levels is the major concern in assessing the minimum drainage disposal requirements. When aimed at maximising source reduction or reuse of drainage water, minimum leaching is required to maintain favourable salt balances in the rootzone. The minimum LF depends on the salinity of the irrigation water and the salt tolerances of the crops grown.

Minimum disposal requirements also depend on the requirements of downstream water uses. For example, many inland fisheries are currently threatened because of increasing water pollution, degradation of aquatic habitats and excessive water abstraction. With the increase in environmental awareness, there is increasing pressure to preserve sufficient water for aquatic environments, habitats and biodiversity. For example, wetlands are considered to be among the world's most valuable ecological sites. They form the habitat for many species of plants and animals.

Wetlands perform various other vital functions, e.g. water storage, flood mitigation and water purification. Wetlands also provide economic benefits, of which fisheries, recreation and tourism are among the most important. These functions, values and attributes can only be maintained if the ecological processes of wetlands continue functioning. If the ecological processes in the wetlands depend on drainage water inflow and if the functions of the wetlands are to be sustained, the minimum drainage discharge will depend on the quality and quantity of drainage water required for maintaining the ecological processes.

Depending on the location, hydrology and topography of the drainage basin (and the ecology and environmental conditions of receiving water bodies), drainage water might be disposed to open surface water bodies, e.g. rivers, lakes, outfall drains, and oceans. The oceans are often regarded as the safest and the final disposal site for agricultural drainage water. This is true unless drainage water is contaminated with sediments, nutrients and other pollutants and the disposal site is in the vicinity of fragile coastal ecosystems such as mangroves and coral reefs. Therefore, at the point of discharge, pollution may be of much concern.

In general, oceans have significant dilution or assimilative capacity. However, this is also limited in many cases especially in enclosed and semi-enclosed seas. Inland drainage water disposal to freshwater bodies such as lakes and rivers requires care. Rivers are normally used for different water use purposes requiring certain qualities of water, while the accumulation of salts and other pollutants in freshwater lakes threatens ecosystem functions and aquatic life.

Disposal into rivers or oceans is not always possible. In closed drainage basins, alternative disposal options need to be sought. The options for closed hydrologic basins include evaporation ponds, deep-well injection and integrated drainage management systems. Drainage water management in closed basins presents numerous environmental and water quality challenges.

Disposal in wetlands offers much potential but at the same time there are also accompanying constraints. Wetlands might be either natural or specially constructed for drainage water reuse. In the former case, the maintenance of ecological functions depends on the quality and quantity of drainage water inflow.

Disposal in Freshwater Bodies

The main aim of safe disposal in freshwater bodies such as lakes and rivers is to protect beneficial downstream water uses. Rivers and lakes are normally multifunctional (municipal drinking-water, industrial water, fishing, recreation and agriculture) and have an intrinsic ecological value. Furthermore, rivers feed lakes, floodplains, wetlands, estuaries and bays. To protect these functions, it is necessary to determine the assimilative capacity of the river and, where necessary, of connected or receiving ecosystems. It is also necessary to identify the constituents of concern in the drainage water in order to determine its discharge requirements. Moreover, the effects on sediments, riparian habitats and floodplains warrant consideration.

The discharge requirements should specify the maximum allowable concentration of each constituent of concern and the volume of drainage water that will be acceptable. Concentration is the water quality parameter normally used in drinking-water standards or for the health of aquatic animals and plants. However, placing only concentration limits on discharge might encourage dilution and inefficient water use.

Where both concentration and load limits (product of water concentration and water volume) are enforced, they tend to promote efficient water use. For example, in the San Joaquin Valley, the discharge limitations into the San Joaquin River include limits on concentration of specific constituents and their loads. The concentration limitations include limits on salinity, boron, selenium and molybdenum in order to protect downstream water quality for domestic and agricultural uses and to safeguard the environment. To discourage dilution or inefficient water usage, load limits for these constituents are enforced for the discharge of irrigation return flows into the San Joaquin River.

The assimilative capacity of the receiving water body varies from place to place and from time to time depending on numerous local conditions. These include climate and physical conditions and the upstream uses. The River Yamuna in Haryana, India, provides an example of seasonal differences in disposal opportunities. During the dry winter season, November-March, the flow of the River Yamuna is less than 50 m^3/s. From July to September, during the monsoon season, the flow exceeds 1 000 m^3/s with a peak of 1 150 m^3/s in August. The salinity of river water during monsoon is less than 0.2 dS/m. The river's high flow and low salinity during the monsoon period provides an opportunity for the disposal of saline effluents.

As the water table shows a marked rise during the monsoon season, major pumping for drainage takes place between July and September. Table 3 shows projections of the amounts of subsurface drainage water that could be disposed into the River Yamuna. The criterion used was that the resultant salinity of the water in the river after mixing should be less than 0.75 dS/m. A significant drainable area with salinity problems can use the River Yamuna as a possible drainage outlet between July and September.

Table 3. Allowable subsurface drainage discharge and drainable area into the River Yamuna

Month	Allowable discharge(m^3/s)		Drainable area(ha)	
	Effluent salinity (dS/m)			
	6	10	6	10
June	0.9	0.5	5 000	3 000
July	25.4	14.4	146 000	83 000
August	47.6	27.0	274 000	156 000
September	6.5	3.7	37 000	21 000
October	3.0	1.7	17 000	10 000

Drainage from the Grassland subarea in the San Joaquin Valley provides another example of the possibilities and constraints of river disposal. In the recent past, irrigated agriculture discharged its irrigation return flow into the North and South Grasslands District, an entity devoted to wetland habitat for private duck clubs and birdwatching as well as pastureland for grazing animals. Since the discovery of selenium poisoning of waterbirds at Kesterson Reservoir from subsurface drainage waters conveyed from the Westlands Water District, drainage from the problem area is no longer discharged into the Grasslands Water District. Instead, it is discharged directly into the San Joaquin River at Mud Slough via the San Luis Drain and a bypass.

Although their flows are minor, Salt Slough and Mud Slough are major contributors of TDS and selenium in the lower reaches of the San Joaquin River. Thus, the Central Valley Regional Water Quality Control Board (CVRWQCB) has placed monthly waste discharge requirements for Salt Slough and Mud Slough. The water quality objectives were established to protect downstream water uses as well as export water to southern California. The salinity water quality objective for Vernalis, a benchmark station on the river, is 1 dS/m for a 30-day running average from 1 September to 30 March, and 0.7 dS/m for a 30-day running average from 1 April to 31 August.

The boron objective is placed from Sack Dam to Merced River, the reach above and below Salt Slough and Mud Slough, at 2.0 mg/litre monthly mean for the year with a maximum of 5.8 mg/litre for a given month. The selenium water quality objective is placed on Salt Slough and Mud Slough at 5 mg/litre, 4-day average, to protect aquatic biota and a maximum annual load of 3 632 kg to control mass emission. Exceeding the specified allowable loads of salt and selenium results in monetary fines to the dischargers. The selenium load is the most difficult for irrigated agriculture to meet.

Generally, rivers continually cleanse themselves. In most cases, lakes do not have this capacity as they may not have an outlet, or flow volume from the lake is limited. The disposal of drainage water into lakes might cause substantial long-term problems. Thus, it involves special considerations. There are numerous examples from all over the world where disposal of agricultural drainage water into freshwater lakes, often in combination with reduced freshwater inflow, has added to the development of immense and catastrophic environmental problems. Examples include Lake Chapala in Mexico, Lake Manchhar in Pakistan, Lake Biwa in Japan, and other lakes where significant effort is required to halt environmental degradation.

Disposal into Evaporation Ponds

In inland drainage basins without an adequate outlet to a river, lake or sea, or where disposal restrictions prevent discharge into rivers and lakes, one of the few options is disposal of drainage effluent into constructed evaporation ponds or natural depressions. Discharge into natural depressions has been practised for centuries. The impounded water is dissipated by evaporation, transpiration and seepage losses. Disposal into constructed ponds is practised worldwide. The planning and design of evaporation ponds have to take account of numerous environmental problems. These include:

— waterlogging and salinisation problems in adjacent areas resulting from excessive seepage losses;

— salt-dust and sprays to areas downwind of the pond surface during dry and windy periods, which might damage vegetation and affect the health of humans and animals; and

— concentration of trace elements that might become toxic to fish and waterbirds because of bioaccumulation in the aquatic food chain.

The following examples from Pakistan, the United States of America and Australia highlight the opportunities and constraints of evaporation ponds.

In Pakistan, evaporation ponds have been constructed under the Salinity Control and Reclamation Project (SCARP) VI for the disposal of the highly saline drainage effluent from the irrigated areas on the fringes of the desert located 500-800 km from the sea. SCARP VI has employed vertical drainage to lower and maintain the water table at a depth of 2.3 m over an area of 152 000 ha. The area is underlain by groundwater with an EC ranging from 15 to 34 dS/m. A total of 514 tubewells, ranging in capacity from 30 to 90 litres/s, have been installed to pump out 600 million m^3 annually. This water is discharged into open surface drains for conveyance to the evaporation ponds for final disposal. Allowing for conveyance losses, the drainable surplus to be disposed of is an estimated 540 million m^3. It was anticipated that the resultant salinity in the drainage water would be of the order of 30 dS/m. The design of the evaporation ponds assumed a percolation rate of 280 mm per year (half of the estimated value) and an evaporation rate of 1 700 mm per year (against the observed value of 1 800 mm). The surface area of the ponds was estimated to be 27 300 ha. The evaporation ponds were developed in a desert area on the fringes of the Indus Plain just beyond the canal command area of the project. The pond area is characterised by interdunal depressions (with highly sodic soils), between longitudinal sand dunes 4-9 m high. The underlying groundwater is highly brackish.

For the development of the evaporation ponds, dykes were provided across the saddles and channels cut across the dunes to form a series of interconnected ponds. The ponds that have been developed so far have a surface area of 13 350 ha and a distance of 1-3 km separates their edges from the irrigated area. The drainage effluent is discharged into the evaporation ponds by gravity flow from one drainage system and pumped from another drainage system further downslope.

The operation of the ponds started in 1989 and by May 1998 it was reported that water had spread over a pond surface area of 4 200 ha. The period of operation has been too short to establish a state of equilibrium. However, soon after ponding started some irrigated areas close to the ponds were severely affected by waterlogging. This could be ascribed to significant seepage losses from the ponds generating a zone of high groundwater, obstructing or retarding the natural subsurface drainage from the irrigated lands. The seepage affected area had grown to about 4 000 ha in 1993, even though the tubewell pumping was effective in lowering the water table to 3 m in parts of the drained area.

In California, between 1972 and 1985, 28 evaporation ponds were constructed covering an area of about 2 800 ha receiving about 39 million m^3 annually of subsurface drainage from 22 700 ha of tile drained fields. Basinwide, the pond area is about 12 percent of the total area drained. Most of the ponds are located in the Tulare subarea, a closed basin in the San Joaquin Valley. The salt concentration in the waters discharged into these ponds ranged from 6 to 70 dS/m with an annual salt load of 0.88 million tonnes, about 25 percent of the annual salt load accumulating in the more than 0.9 million ha of cropland.

The concentration range of selenium, the principal constituent of concern, is from 1 to over 600 ppb. A selenium concentration of 2 ppb is considered the upper limit for aquatic life in ponds as selenium tends to bioaccumulate 1 000- to 2 500-fold in the aquatic food chain. Selenium is toxic to waterbirds as it substitutes for sulphur in essential amino acids resulting in embryo deformities and reduced reproduction rates. Ten ponds are active and managed by seven operators. The other ponds have been voluntarily deactivated due to the high costs of meeting the waste discharge requirements and mitigation measures for bird toxicity. The state/regional water quality regulatory agency ordered the closure of several ponds because of the unacceptable toxic effects of selenium on waterbirds from selenium present in the impounded drainage waters.

The waste discharge requirements for pond disposal define the requirements and compliance schedules designed to discourage wildlife use of evaporation ponds and/or provide mitigation and compensation measures to offset adverse effects on birds.

The evaporation ponds are constructed by excavating the soil to form berms with side-slopes of at least 3:1 (h:v). The pond bottoms are unlined but compacted, the pond environs are kept free of vegetation, and a minimum water depth is maintained at 60 cm. Migratory waterbirds attracted to the pond site are to be scared off, and bird diseases kept under control.

Compliance monitoring includes seasonal drainage water and sediment selenium concentrations, and biological monitoring of birds for abundance and symptoms of toxicity. When adverse selenium impacts are noted, off-site mitigation measures must be implemented with either compensation or alternative habitats as guided by approved protocols and risk analysis methods.

Compensation habitats are constructed to compensate for unavoidable migratory bird losses by providing a wetland habitat safe from selenium and predators.

Alternative habitats are year-round freshwater habitats immediately adjacent to contaminated ponds in order to provide dietary dilution to selenium exposure. In spite of the stringent waste discharge requirements and associated high costs, farmers and districts in the Tulare subarea use evaporation ponds for drainage water disposal basins. This is because there are no opportunities for off-site discharge into the San Joaquin River or other designated sinks for drainage waters. Evaporation ponds are the only economic means of disposal of saline drainage in the Tulare-Kern subareas.

Treatment of Drainage Effluent

Treating drainage water is normally one of the last drainage water management options to be considered. This is due to the high costs involved and to uncertainty about the treatment level achievable. The treatment of drainage water should be considered where all other drainage water management measures fail to guarantee safe disposal or where it is financially attractive. For subsurface drainage water containing very high levels of salinity, selenium and other trace elements, the treatment objectives are:

i) reduce salts and toxic constituents below hazardous levels;

ii) meet agricultural water management goals;

iii) meet water quality objectives in surface waters; and

iv) reduce constituent levels below risk levels for wildlife.

The treatment of agricultural drainage water presents a challenge due to the complex chemical characteristics of most drainage waters. Table 4 details the average chemical quality of subsurface drainage waters disposed into Kesterson Reservoir in the San Joaquin Valley as well as those disposed into evaporation ponds.

The drainage waters are saline and of the $NaCl$-Na_2SO_4-type water. The waters conveyed by the San Luis Drain came from a single site in Westlands Water District in contrast to the evaporation pond waters that came from 27 sites. There are numerous wastewater treatment processes for industrial and urban wastewater and for the preparation of drinking-water. Many of them offer potential for the treatment of agricultural drainage water.

Treatment processes for drainage water can be divided into processes that reduce the total salinity of the drainage water and processes that remove specific ions. Methods for the removal of trace elements can be biological, physical and chemical.

Table 4. Average composition of agricultural tile drainage water in the San Luis Drain (drainage waters disposed into evaporation basins in parenthesis)

Constituent	Concentration (ppm)	Constituent	Concentration (ppb)
Sodium	2 230	Boron	14 400 (25 000)
Calcium	554	Selenium	325 (16)
Magnesium	270	Arsenic	1 (101)
Potassium	6	Molybdenum	ND (2 817)
Alkalinity as $CaCO_3$	196	Uranium	ND (308)
Sulphate	4 730	Vanadium	ND (22)
Chloride	1 480	Strontium	6 400
Nitrate	48	Total chromium	19
Silica	37	Cadmium	<1
TDS	9 820 (31 000)	Copper	4
Suspended solids	11	Lead	3
Total organic carbon	10.2	Manganese	25
COD	32	Iron	110
BOD	3.2	Mercury	<0.1
		Nickel	14
		Zinc	33

Desalinisation

There are numerous desalinisation processes including ion exchange, distillation, electrodialysis and reverse osmosis. Of these processes, reverse osmosis is considered to be the most promising for the treatment of agricultural drainage water mainly due to its comparatively low cost.

Reverse osmosis is a process capable of removing different contaminants including dissolved salts and organics. In reverse osmosis, a semi-permeable membrane separates water from dissolved salts and other suspended solids. Pressure is applied to the feed-water, forcing the water through the membrane leaving behind salts and suspended materials in a brine stream. The energy consumption of the process depends on the salt concentration of the feed-water and the salt concentration of the effluent. Depending on the quality of the water to be treated, pretreatment might be crucial to preventing fouling of the membrane.

References

Allen, RG,L.S. Pereira, Dr. Raes, M. Smith, *Crop Evapotranspiration (Guidelines for Computing Crop Water Requirement).*, FAO Irrigation and Drainage Paper No.56. 1998.

Melden , D., *Water for food, water for life; a Comprehensive Assessment of Water Management in Agriculture;* International Water Management Institute (IWMI). 2007.

Redzeweski, J.R, S. Nairizi, I*rrigation Planning Based on Water Deficits*; Water Resources Bulletin American Water Resources association, 1979.

Nairizi, I,R. Redzeweski, *Effects of Dated Soil Moisture Stress on Crop Yields;* Irrigation Engineering Research Group, Department of Civil Engineering University of Southampton, 1976.

UNESCO; *Water in a Changing World,* the United Nation World Water Development Report. 2009.

8

Wastewater Treatment and Use in Agriculture

In many arid and semi-arid countries water is becoming an increasingly scarce resource and planners are forced to consider any sources of water which might be used economically and effectively to promote further development. At the same time, with population expanding at a high rate, the need for increased food production is apparent. The potential for irrigation to raise both agricultural productivity and the living standards of the rural poor has long been recognized. Irrigated agriculture occupies approximately 17 percent of the world's total arable land but the production from this land comprises about 34 percent of the world total. This potential is even more pronounced in arid areas, such as the Near East Region, where only 30 percent of the cultivated area is irrigated but it produces about 75 percent of the total agricultural production. In this same region, more than 50 percent of the food requirements are imported and the rate of increase in demand for food exceeds the rate of increase in agricultural production.

Whenever good quality water is scarce, water of marginal quality will have to be considered for use in agriculture. Although there is no universal definition of 'marginal quality' water, for all practical purposes it can be defined as water that possesses certain characteristics which have the potential to cause problems when it is used for an intended purpose. For example, brackish water is a marginal quality water for agricultural use because of its high dissolved salt content, and municipal wastewater is a marginal quality water because of the associated health hazards. From the viewpoint of irrigation, use of a 'marginal' quality water requires more complex management practices and more stringent monitoring procedures than when good quality water is used. This publication deals with agricultural use of municipal wastewater, which is primarily domestic sewage but possibly contains a proportion of industrial effluents discharged to public sewers.

Expansion of urban populations and increased coverage of domestic water supply and sewerage give rise to greater quantities of municipal wastewater. With the current emphasis on environmental health and water pollution issues, there is an increasing awareness of the need to dispose of these wastewaters safely and beneficially. Use of wastewater in agriculture could be an important consideration when its disposal is being planned in arid and semi-arid regions. However it should be realized that the quantity of wastewater available in most countries will account for only a small fraction of the total irrigation water requirements. Nevertheless, wastewater use will result in the conservation of higher quality water and its use for purposes other than irrigation. As the marginal cost of alternative supplies of good quality water will usually be higher in water-short areas, it makes good sense to incorporate agricultural reuse into water resources and land use planning.

Properly planned use of municipal wastewater alleviates surface water pollution problems and not only conserves valuable water resources but also takes advantage of the nutrients contained in sewage to grow crops. The availability of this additional water near population centres will increase the choice of crops which farmers can grow. The nitrogen and phosphorus content of sewage might reduce or eliminate the requirements for commercial fertilizers. It is advantageous to consider effluent reuse at the same time as wastewater collection, treatment and disposal are planned so that sewerage system design can be optimized in terms of effluent transport and treatment methods. The cost of transmission of effluent from inappropriately sited sewage treatment plants to distant agricultural land is usually prohibitive. Additionally, sewage treatment techniques for effluent discharge to surface waters may not always be appropriate for agricultural use of the effluent.

Many countries have included wastewater reuse as an important dimension of water resources planning. In the more arid areas of Australia and the USA wastewater is used in agriculture, releasing high quality water supplies for potable use. Some countries, for example the Hashemite

Kingdom of Jordan and the Kingdom of Saudi Arabia, have a national policy to reuse all treated wastewater effluents and have already made considerable progress towards this end. In China, sewage use in agriculture has developed rapidly since 1958 and now over 1.33 million hectares are irrigated with sewage effluent. It is generally accepted that wastewater use in agriculture is justified on agronomic and economic grounds but care must be taken to minimize adverse health and environmental impacts. The purpose of this document is to provide countries with guidelines for wastewater use in agriculture which will allow the practice to be adopted with complete health and environmental security.

CHARACTERISTICS OF WASTEWATERS

Municipal wastewater is mainly comprised of water (99.9%) together with relatively small concentrations of suspended and dissolved organic and inorganic solids. Among the organic substances present in sewage are carbohydrates, lignin, fats, soaps, synthetic detergents, proteins and their decomposition products, as well as various natural and synthetic organic chemicals from the process industries.

Municipal wastewater also contains a variety of inorganic substances from domestic and industrial sources, including a number of potentially toxic elements such as arsenic, cadmium, chromium, copper, lead, mercury, zinc, etc. Even if toxic materials are not present in concentrations likely to affect humans, they might well be at phytotoxic levels, which would limit their agricultural use. However, from the point of view of health, a very important consideration in agricultural use of wastewater, the contaminants of greatest concern are the pathogenic micro- and macro-organisms.

Pathogenic bacteria will be present in wastewater at much lower levels than the coliform group of bacteria, which are much easier to identify and enumerate (as total coliforms/100ml). *Escherichia coli* are the most widely adopted indicator of faecal pollution and they can also be isolated and identified fairly simply, with their numbers usually being given in the form of faecal coliforms (FC)/100 ml of wastewater.

Parameters of Health Significance

Organic chemicals usually exist in municipal wastewaters at very low concentrations and ingestion over prolonged periods would be necessary to produce detrimental effects on human health. This is not likely to occur with agricultural/aquacultural use of wastewater, unless cross-connections with potable supplies occur or agricultural workers are not properly instructed, and can normally be ignored. The principal health hazards associated with the chemical constituents of wastewaters, therefore, arise from the contamination of crops or groundwaters. Hillman has drawn attention to the particular concern attached to the cumulative poisons, principally heavy metals, and carcinogens, mainly organic chemicals.

Pathogenic organisms give rise to the greatest health concern in agricultural use of wastewaters, yet few epidemological studies have established definitive adverse health impacts attributable to the practice. Shuval et al. reported on one of the earliest evidences connecting agricultural wastewater reuse with the occurrence of disease. It would appear that in areas of the world where helminthic diseases caused by *Ascaris* and *Trichuris* spp. are endemic in the population and where raw untreated sewage is used to irrigate salad crops and/or vegetables eaten uncooked, transmission of these infections is likely to occur through the consumption of such crops.

There is only limited evidence indicating that beef tapeworm *(Taenia saginata)* can be transmitted to the population consuming the meat of cattle grazing on wastewater irrigated fields or fed crops from such fields. However, there is strong evidence from Melbourne, Australia and from Denmark that cattle grazing on fields freshly irrigated with raw wastewater, or drinking from raw wastewater canals or ponds, can become heavily infected with the disease (cysticerosis).

Indian studies, reported by Shuval et al. , have shown that sewage farm workers exposed to raw wastewater in areas where*Ancylostoma* (hookworm) and *Ascaris* (nematode) infections are endemic have significantly excess levels of infection with these two parasites compared with other agricultural workers in similar occupations. Furthermore, the studies indicated that the intensity of the Ascaris infections (the number of worms infesting the intestinal tract of an individual) in the sample of sewage farm workers was very much greater than in the control sample. In the case of the hookworm infections, the severity of the health effects was a function of the worm load of individuals, which was found to be related to the degree of exposure and the length of time of exposure to the hookworm larvae. Sewage farm workers are also liable to become infected with cholera if practising irrigation with raw wastewater derived from an urban area in which a cholera epidemic is in progress. Morbidity and serological studies on wastewater irrigation workers or wastewater treatment plant workers occupationally exposed to wastewater directly and to wastewater aerosols have not been able to demonstrate excess prevalence of viral diseases.

Parameters of Agricultural Significance

The quality of irrigation water is of particular importance in arid zones where extremes of temperature and low relative humidity result in high rates of evaporation, with consequent deposition of salt which tends to accumulate in the soil profile. The physical and mechanical properties of the soil, such as dispersion of particles, stability of aggregates, soil structure and permeability, are very sensitive to the type of exchangeable ions present in irrigation water. Thus, when effluent use is being planned, several factors related to soil properties must be taken into consideration.

Another aspect of agricultural concern is the effect of dissolved solids (TDS) in the irrigation water on the growth of plants. Dissolved salts increase the osmotic potential of soil water and an increase in osmotic pressure of the soil solution increases the amount of energy which plants must expend to take up water from the soil. As a result, respiration is increased and the growth and yield of most plants decline progressively as osmotic pressure increases. Although most plants respond to salinity as a function of the total osmotic potential of soil water, some plants are susceptible to specific ion toxicity.

Many of the ions which are harmless or even beneficial at relatively low concentrations may become toxic to plants at high concentration, either through direct interference with metabolic processes or through indirect effects on other nutrients, which might be rendered inaccessible. Morishita has reported that irrigation with nitrogen-enriched polluted water can supply a considerable excess of nutrient nitrogen to growing rice plants and can result in a significant yield loss of rice through lodging, failure to ripen and increased susceptibility to pests and diseases as a result of over-luxuriant growth. He further reported that non-polluted soil, having around 0.4 and 0.5 ppm cadmium, may produce about 0.08 ppm Cd in brown rice, while only a little increase up to 0.82, 1.25 or 2.1 ppm of soil Cd has the potential to produce heavily polluted brown rice with 1.0 ppm Cd.

Important agricultural water quality parameters include a number of specific properties of water that are relevant in relation to the yield and quality crops, maintenance of soil productivity and protection of the environment. These parameters mainly consist of certain physical and chemical characteristics of the water

Total Salt Concentration

Total salt concentration (for all practical purposes, the total dissolved solids) is one of the most important agricultural water quality parameters. This is because the salinity of the soil water is related to, and often determined by, the salinity of the irrigation water. Accordingly, plant growth, crop yield and quality of produce are affected by the total dissolved salts in the irrigation water. Equally, the rate of accumulation of salts in the soil, or soil salinization, is also directly affected by the salinity of the irrigation water. Total salt concentration is expressed in milligrams per litre (mg/l) or parts per million (ppm).

Electrical Conductivity

Electrical conductivity is widely used to indicate the total ionized constituents of water. It is directly related to the sum of the cations (or anions), as determined chemically and is closely correlated, in general, with the total salt concentration. Electrical conductivity is a rapid and reasonably precise determination and values are always expressed at a standard temperature of 25°C to enable comparison of readings taken under varying climatic conditions. It should be noted that the electrical conductivity of solutions increases approximately 2 percent per °C increase in temperature. In this publication, the symbol EC_w, is used to represent the electrical conductivity of irrigation water and the symbol EC_e is used to designate the electrical conductivity of the soil saturation extract. The unit of electrical conductivity is deciSiemen per metre (dS/m).

Sodium Adsorption Ratio

Sodium is an unique cation because of its effect on soil. When present in the soil in exchangeable form, it causes adverse physico-chemical changes in the soil, particularly to soil structure. It has the ability to disperse soil, when present above a certain threshold value, relative to the concentration of total dissolved salts. Dispersion of soils results in reduced infiltration rates of water and air into the soil. When dried, dispersed soil forms crusts which are hard to till and interfere with germination and seedling emergence. Irrigation water could be a source of excess sodium in the soil solution and hence it should be evaluated for this hazard.

Toxic Ions

Irrigation water that contains certain ions at concentrations above threshold values can cause plant toxicity problems. Toxicity normally results in impaired growth, reduced yield, changes in the morphology of the plant and even its death. The degree of damage depends on the crop, its stage of growth, the concentration of the toxic ion, climate and soil conditions.

The most common phytotoxic ions that may be present in municipal sewage and treated effluents in concentrations such as to cause toxicity are: boron (B), chloride (Cl) and sodium (Na). Hence, the concentration of these ions will have to be determined to assess the suitability of waste-water quality for use in agriculture.

Trace Elements and Heavy Metals

A number of elements are normally present in relatively low concentrations, usually less than a few mg/l, in conventional irrigation waters and are called trace elements. They are not normally included in routine analysis of regular irrigation water, but attention should be paid to them when using sewage effluents, particularly if contamination with industrial wastewater discharges is suspected. These include Aluminium (A1), Beryllium (Be), Cobalt (Co), Fluoride (F), Iron (Fe), Lithium (Li), Manganese (Mn), Molybdenum (Mo), Selenium (Se), Tin (Sn), Titanium (Ti), Tungsten (W) and Vanadium (V). Heavy metals are a special group of trace elements which have been shown to create definite health hazards when taken up by plants. Under this group are included, Arsenic (As), Cadmium (Cd), Chromium (Cr), Copper (Cu), Lead (Pb), Mercury (Hg) and Zinc (Zn). These are called heavy metals because in their metallic form, their densities are greater than 4g/cc.

pH

pH is an indicator of the acidity or basicity of water but is seldom a problem by itself. The normal pH range for irrigation water is from 6.5 to 8.4; pH values outside this range are a good warning that the water is abnormal in quality. Normally, pH is a routine measurement in irrigation water quality assessment.

Wastewater Treatment

The principal objective of wastewater treatment is generally to allow human and industrial effluents to be disposed of without danger to human health or unacceptable damage to the natural environment. Irrigation with wastewater is both disposal and utilization and indeed is an effective form of wastewater disposal (as in slow-rate land treatment). However, some degree of treatment must normally be provided to raw municipal wastewater before it can be used for agricultural or landscape irrigation or for aquaculture. The quality of treated effluent used in agriculture has a great influence on the operation and performance of the wastewater-soil-plant or aquaculture system. In the case of irrigation, the required quality of effluent will depend on the crop or crops to be irrigated, the soil conditions and the system of effluent distribution adopted. Through crop restriction and selection of irrigation systems which minimize health risk, the degree of pre-application wastewater treatment can be reduced. A similar approach is not feasible in aquaculture systems and more reliance will have to be placed on control through wastewater treatment.

The most appropriate wastewater treatment to be applied before effluent use in agriculture is that which will produce an effluent meeting the recommended microbiological and chemical quality guidelines both at low cost and with minimal operational and maintenance requirements (Arar 1988). Adopting as low a level of treatment as possible is especially desirable in developing countries, not only from the point of view of cost but also in acknowledgement of the difficulty of operating complex systems reliably. In many locations it will be better to design the reuse system to accept a low-grade of effluent rather than to rely on advanced treatment processes producing a reclaimed effluent which continuously meets a stringent quality standard.

Nevertheless, there are locations where a higher-grade effluent will be necessary and it is essential that information on the performance of a wide range of wastewater treatment technology should be available. The design of wastewater treatment plants is usually based on the need to reduce organic and suspended solids loads to limit pollution of the environment. Pathogen removal has very rarely been considered an objective but, for reuse of effluents in agriculture, this must now be of primary concern and processes should be selected and designed accordingly. Treatment to remove wastewater constituents that may be toxic or harmful to crops, aquatic plants (macrophytes) and fish is technically possible but is not normally economically feasible. Unfortunately, few performance data on wastewater treatment plants in developing countries are available and even then they do not normally include effluent quality parameters of importance in agricultural use.

The short-term variations in wastewater flows observed at municipal wastewater treatment plants follow a diurnal pattern. Flow is typically low during the early morning hours, when water consumption is lowest and when the base flow consists of infiltration-inflow and small quantities of sanitary wastewater. A first peak of flow generally occurs in the late morning, when wastewater from the peak morning water use reaches the treatment plant, and a second peak flow usually occurs in the evening. The relative magnitude of the peaks and the times at which they occur vary from country to country and with the size of the community and the length of the sewers. Small communities with small sewer systems have a much higher ratio of peak flow to average flow than do large communities. Although the magnitude of peaks is attenuated as wastewater passes through a treatment plant, the daily variations in flow from a municipal treatment plant make it impracticable, in most cases, to irrigate with effluent directly from the treatment plant. Some form of flow equalization or short-term storage of treated effluent is necessary to provide a relatively constant supply of reclaimed water for efficient irrigation, although additional benefits result from storage.

Conventional Wastewater Treatment Processes

Conventional wastewater treatment consists of a combination of physical, chemical, and biological processes and operations to remove solids, organic matter and, sometimes, nutrients from wastewater. General terms used to describe different degrees of treatment, in order of increasing treatment level, are preliminary, primary, secondary, and tertiary and/or advanced wastewater treatment. In some countries, disinfection to remove pathogens sometimes follows the last treatment step.

Preliminary treatment

The objective of preliminary treatment is the removal of coarse solids and other large materials often found in raw wastewater. Removal of these materials is necessary to enhance the operation and maintenance of subsequent treatment units. Preliminary treatment operations typically include coarse screening, grit removal and, in some cases, comminution of large objects. In grit chambers, the velocity of the water through the chamber is maintained sufficiently high, or air is used, so as to prevent the settling of most organic solids. Grit removal is not included as a preliminary treatment step in most small wastewater treatment plants. Comminutors are sometimes adopted to supplement coarse screening and serve to reduce the size of large particles so that they will be removed in the form of a sludge in subsequent treatment processes. Flow measurement devices, often standing-wave flumes, are always included at the preliminary treatment stage.

Primary treatment

The objective of primary treatment is the removal of settleable organic and inorganic solids by sedimentation, and the removal of materials that will float (scum) by skimming. Approximately 25 to 50% of the incoming biochemical oxygen demand (BOD_5), 50 to 70% of the total suspended solids (SS), and 65% of the oil and grease are removed during primary treatment. Some organic nitrogen, organic phosphorus, and heavy metals associated with solids are also removed during primary sedimentation but colloidal and dissolved constituents are not affected. The effluent from primary sedimentation units is referred to as primary effluent.

In many industrialized countries, primary treatment is the minimum level of preapplication treatment required for wastewater irrigation. It may be considered sufficient treatment if the wastewater is used to irrigate crops that are not consumed by humans or to irrigate orchards, vineyards, and some processed food crops. However, to prevent potential nuisance conditions in storage or flow-equalizing reservoirs, some form of secondary treatment is normally required in these countries, even in the case of non-food crop irrigation. It may be possible to use at least a portion of primary effluent for irrigation if off-line storage is provided.

Primary sedimentation tanks or clarifiers may be round or rectangular basins, typically 3 to 5 m deep, with hydraulic retention time between 2 and 3 hours. Settled solids (primary sludge) are normally removed from the bottom of tanks by sludge rakes that scrape the sludge to a central well from which it is pumped to sludge processing units. Scum is swept across the tank surface by water jets or mechanical means from which it is also pumped to sludge processing units.

In large sewage treatment plants (> 7600 m^3/d in the US), primary sludge is most commonly processed biologically by anaerobic digestion. In the digestion process, anaerobic and facultative bacteria metabolize the organic material in sludge, thereby reducing the volume requiring ultimate disposal, making the sludge stable (nonputrescible) and improving its dewatering characteristics. Digestion is carried out in covered tanks (anaerobic digesters), typically 7 to 14 m deep. The residence time in a digester may vary from a minimum of about 10 days for high-rate digesters (well-mixed and heated) to 60 days or more in standard-rate digesters. Gas containing about 60 to 65% methane is produced during digestion and can be recovered as an energy source. In small sewage treatment plants, sludge is processed in a variety of ways including: aerobic digestion, storage in sludge lagoons, direct application to sludge drying beds, in-process storage (as in stabilization ponds), and land application.

Secondary treatment

The objective of secondary treatment is the further treatment of the effluent from primary treatment to remove the residual organics and suspended solids. In most cases, secondary treatment follows primary treatment and involves the removal of biodegradable dissolved and colloidal organic matter using aerobic biological treatment processes. Aerobic biological treatment is performed in the presence of oxygen by aerobic microorganisms (principally bacteria) that metabolize the organic matter in the wastewater, thereby producing more microorganisms and inorganic end-products (principally CO_2, NH_3, and H_2O). Several aerobic biological processes are used for secondary treatment differing primarily in the manner in which oxygen is supplied to the microorganisms and in the rate at which organisms metabolize the organic matter.

High-rate biological processes are characterized by relatively small reactor volumes and high concentrations of microorganisms compared with low rate processes. Consequently, the growth rate of new organisms is much greater in high-rate systems because of the well controlled environment. The microorganisms must be separated from the treated wastewater by sedimentation to produce clarified secondary effluent. The sedimentation tanks used in secondary treatment, often referred to as secondary clarifiers, operate in the same basic manner as the primary clarifiers described previously. The biological solids removed during secondary sedimentation, called secondary or biological sludge, are normally combined with primary sludge for sludge processing.

Common high-rate processes include the activated sludge processes, trickling filters or biofilters, oxidation ditches, and rotating biological contactors (RBC). A combination of two of these processes in series (e.g., biofilter followed by activated sludge) is sometimes used to treat municipal wastewater containing a high concentration of organic material from industrial sources.

Tertiary and/or advanced treatment

Tertiary and/or advanced wastewater treatment is employed when specific wastewater constituents which cannot be removed by secondary treatment must be removed. Individual treatment processes are necessary to remove nitrogen, phosphorus, additional suspended solids, refractory organics, heavy metals and dissolved solids. Because advanced treatment usually follows high-rate secondary treatment, it is sometimes referred to as tertiary treatment. However, advanced treatment processes are sometimes combined with primary or secondary treatment (e.g., chemical addition to primary clarifiers or aeration basins to remove phosphorus) or used in place of secondary treatment (e.g., overland flow treatment of primary effluent).

An adaptation of the activated sludge process is often used to remove nitrogen and phosphorus and an example of this approach is the 23 Ml/d treatment plant commissioned in 1982 in British Columbia, Canada. Effluent from primary clarifiers flows to the biological reactor, which is physically divided into five zones by baffles and weirs. In sequence these zones are: (i) anaerobic fermentation zone (characterized by very low dissolved oxygen levels and the absence of nitrates); (ii) anoxic zone (low dissolved oxygen levels but nitrates present); (iii) aerobic zone (aerated); (iv) secondary anoxic zone; and (v) final aeration zone. The function of the first zone is to condition the group of bacteria responsible for phosphorus removal by stressing them under low oxidation-reduction conditions, which results in a release of phosphorus equilibrium in the cells of the bacteria. On subsequent exposure to an adequate supply of oxygen and phosphorus in the aerated zones, these cells rapidly accumulate phosphorus considerably in excess of their normal metabolic requirements. Phosphorus is removed from the system with the waste activated sludge.

Most of the nitrogen in the influent is in the ammonia form, and this passes through the first two zones virtually unaltered. In the third aerobic zone, the sludge age is such that almost complete nitrification takes place, and the ammonia nitrogen is converted to nitrites and then to nitrates. The nitrate-rich mixed liquor is then recycled from the aerobic zone back to the first anoxic zone. Here denitrification occurs, where the recycled nitrates, in the absence of dissolved oxygen, are reduced by facultative bacteria to nitrogen gas, using the influent organic carbon compounds as hydrogen donors. The nitrogen gas merely escapes to atmosphere. In the second anoxic zone, those nitrates which were not recycled are reduced by the endogenous respiration of bacteria. In the final re-aeration zone, dissolved oxygen levels are again raised to prevent further denitrification, which would impair settling in the secondary clarifiers to which the mixed liquor then flows.

An experimentation programme on this plant demonstrated the importance of the addition of volatile fatty acids to the anaerobic fermentation zone to achieve good phosphorus removal. These essential short-chain organics (mainly acetates) are produced by the controlled fermentation of primary sludge in a gravity thickener and are released into the thickener supernatent, which can be fed to the head of the biological reactor. Without this supernatent return flow, overall phosphorus removal quickly dropped to levels found in conventional activated sludge plants. Performance data over three years have proved that, with thickener supernatent recycle, effluent quality median values of 0.5-1.38 mg/l Ortho-P, 1.4-1.6 mg/l Total nitrogen and 1.4-2.0 mg/l nitrate-N are achievable. This advanced biological wastewater treatment plant cost only marginally more than a conventional activated sludge plant but nevertheless involved considerable investment. Furthermore, the complexity of the process and the skilled operation required to achieve consistent results make this approach unsuitable for developing countries.

In many situations, where the risk of public exposure to the reclaimed water or residual constituents is high, the intent of the treatment is to minimize the probability of human exposure to enteric viruses and other pathogens. Effective disinfection of viruses is believed to be inhibited by suspended and colloidal solids in the water, therefore these solids must be removed by advanced treatment before the disinfection step. The sequence of treatment often specified in the United States is: secondary treatment followed by chemical coagulation, sedimentation, filtration, and disinfection.

Disinfection

Disinfection normally involves the injection of a chlorine solution at the head end of a chlorine contact basin. The chlorine dosage depends upon the strength of the wastewater and other factors, but dosages of 5 to 15 mg/l are common. Ozone and ultra violet (uv) irradiation can also be used for disinfection but these methods of disinfection are not in common use. Chlorine contact basins are usually rectangular channels, with baffles to prevent short-circuiting, designed to provide a contact time of about 30 minutes. However, to meet advanced wastewater treatment requirements, a chlorine contact time of as long as 120 minutes is sometimes required for specific irrigation uses of reclaimed wastewater. The bactericidal effects of chlorine and other disinfectants are dependent upon pH, contact time, organic content, and effluent temperature.

Effluent storage

Although not considered a step in the treatment process, a storage facility is, in most cases, a critical link between the wastewater treatment plant and the irrigation system. Storage is needed for the following reasons:

i. To equalize daily variations in flow from the treatment plant and to store excess when average wastwater flow exceeds irrigation demands; includes winter storage.

ii. To meet peak irrigation demands in excess of the average wastewater flow.

iii. To minimize the effects of disruptions in the operations of the treatment plant and irrigation system. Storage is used to provide insurance against the possibility of unsuitable reclaimed wastewater entering the irrigation system and to provide additional time to resolve temporary water quality problems.

Reliability of conventional and advanced wastewater treatment

Wastewater reclamation and reuse systems should contain both design and operational requirements necessary to ensure reliability of treatment. Reliability features such as alarm systems, standby power supplies, treatment process duplications, emergency storage or disposal of inadequately treated wastewater, monitoring devices, and automatic controllers are important. From a public health standpoint, provisions for adequate and reliabile disinfection are the most essential features of the advanced wastewater treatment process. Where disinfection is required, several reliability features must be incorporated into the system to ensure uninterrupted chlorine feed.

Natural biological treatment systems

Natural low-rate biological treatment systems are available for the treatment of organic wastewaters such as municipal sewage and tend to be lower in cost and less sophisticated in operation and maintenance. Although such processes tend to be land intensive by comparison with the conventional high-rate biological processes already described, they are often more effective in removing pathogens and do so reliably and continuously if properly designed and not overloaded. Among the natural biological treatment systems available, stabilization ponds and land treatment have been used widely around the world and a considerable record of experience and design practice has been documented. The nutrient film technique is a fairly recent development of the hydroponic plant growth system with application in the treatment and use of wastewater.

Treated Wastewater Used in Irrigated Agriculture

In many developing countries, untreated or partially treated waste-water is used to irrigate the cities' own food, fodder, and green spaces. Farmers have been using untreated waste-water for centuries, but greater numbers now depend on it for their livelihoods and this demand has ushered in a range of new waste-water use practices. The use of urban waste-water in agriculture is a centuries-old practice that is receiving renewed attention with the increasing scarcity of freshwater resources in many arid and semiarid regions.

Driven by rapid urbanisation, waste-water is widely used as a low-cost alternative to conventional irrigation water; it supports livelihoods and generates considerable value in urban and peri-urban agriculture despite the health and environmental risks associated with this practice. Though pervasive, this practice is largely unregulated in low-income countries, and the costs and benefits are poorly understood.

Sewage collection and its disposal as waste-water are increasing in developing-country cities as a function of the growth in urban water supply. Increases in urban water supply depend on myriad factors and will likely be unable to keep pace with urban population growth, implying falling per capita water supply rates. In spite of the fact that trends show that rates of urbanisation are likely to slow down in developed countries, in many countries of the developing world urbanisation will continue rapidly. As a result, waste-water flows will increase in the future. In developing countries where investments in water supply far outpace those in sanitation and waste management, suffice it to say that treatment and disposal of waste-water are inadequate or non-existent and that raw sewage - full-strength or diluted - is used and even competed for in order to irrigate food, fodder, ornamental and other crops.

Raw waste-water use in agriculture is presently increasing at close to the rate of urban growth in developing countries subject to urban and peri-urban land being available. Developed countries' populations are expected to decline 6% by 2050, while the global rural population should plateau at approximately 3.2 billion. The result is that after 2015, all worldwide growth in population will take place in developing-country cities. Cities are home to political and economic power and will continue to ensure that their water supply needs are met on a priority basis subject to physical and economic scarcity constraints.

The Millennium Development Goals call for halving the proportion of people without access to improved sanitation or water by 2015. As a result, an additional 1.6 billion people will require access to a water supply - 1.018 billion in urban areas and 581 million in rural areas. Water supply ensures waste-water because the depleted fraction of domestic and residential water use is typically only 15-25% with the remainder returning as waste-water. Although the numbers of urban dwellers in developing countries that continue to rely on septic tanks, cesspits, etc. is unexpectedly high, growing numbers are connected to sewers that deliver waste-water - largely untreated - to downstream areas.

In water-scarce and even humid regions, farmers prize the water and nutrient value and supply reliability of the waste-water stream. And under the most common scenario in water-scare countries of a city expanding more rapidly than its water supply, sewage may water what little green space remains.

Irrigation with untreated waste-water can represent a major threat to public health (of both humans, and livestock), food safety, and environmental quality. The microbial quality of waste-water is usually measured by the concentration of the two primary sources of water-borne infection - faecal coliforms and nematode eggs. A range of viruses and protozoa pose additional health risks. Waste-water has been implicated as an important source of health risk for chronic, low-grade gastrointestinal disease as well as outbreaks of more acute diseases including cholera and typhoid.

Disease agents are found in waste-water that drains from planned residential areas and slums alike. The health of the urban poor is particularly linked to inadequate management of waste-water. Chronic diarrhoeal and gastro-intestinal diseases, which disproportionately affect urban slum dwellers who have inadequate sewerage and sanitation facilities, are clearly major negative outcomes of exposure to waste-water. A primary exposure route for the urban population in general is the consumption of raw vegetables that have been irrigated with waste-water. Additional exposure routes for the urban poor, who are often migrants with little access to health services, include direct contact with solid waste and waste-water, as for instance through riverside open defecation grounds.

Additionally waste-water irrigation of vegetables and fodder may serve as the transmission route for heavy metals in the human food chain. Particularly in South Asia, where per capita milk consumption is the highest in the developing world and growing rapidly, waste-water is increasingly used to irrigate fodder that supplies an urban and peri-urban livestock-based production chain. Evidence of heavy metal transmission through milk is presented by Swarup et al.. In the absence of chilling, storage and transport facilities, milk must be produced as close to market as possible; it represents an important urban and peri-urban agricultural product. Further, fodder cultivation is particularly well matched to waste-water; it requires continual irrigation application and is generally tolerant of the high salinity levels characteristic of urban waste-water.

Finally, the environmental quality of soils, groundwater and surface water, and to a lesser degree, stream channel biota and ecological conditions as indicated by the biodiversity of the waste-water-contaminated river or other receiving water body are often the second-order casualties if waste-water is disposed indiscriminately.

Cities in both arid and humid regions are witnessing unprecedented expansion of urban and peri-urban agriculture using poor-quality water. For example, in Bolivia, indirect use of waste-water takes place in almost all rural and peri-urban areas downstream of the urban centres. Additionally, although waste-water irrigation has been thought to be limited to large cities, in regions such as Gujarat, India, it is common even downstream of small towns and villages.

In sum, waste-water is a resource of growing global importance and its use in agriculture must be carefully managed in order to preserve the substantial benefits while minimising the serious risks. This reality was recognised and its implications deliberated in the Hyderabad Declaration on Waste-water Use in Agriculture, one of the outcomes of a workshop held 11-14 November 2002 in Hyderabad, India and sponsored by the International Water Management Institute (IWMI, based in Colombo, Sri

Lanka) and the International Development Research Centre. Several factors drive waste-water irrigation: the lack of equally remunerative livelihood alternatives, the continued expansion of the waste-water resource base, and the ineffectiveness of regulatory control approaches that have characterised most attempts at management.

The experiences of countries that are in the process or have completed the conversion from untreated to regulated, treated reuse can serve as important lessons. Treated waste-water currently represents approximately 5% of Tunisia's total available water; this is planned to increase to 11% by 2030. Salinity management remains a major objective of the Tunisian waste-water use programme.

In Jordan, waste-water represents 10% of the current total water supply. Groundwater recharge is one of the explicit uses of waste-water in Jordan, but not for aquifers that are used for drinking water supply. The previous (waste-) water quality standards required some revision in order to accommodate Jordan's plans to reuse water, particularly for sprinkler irrigation, which was prohibited for waste-water. In order to meet strict export phytosanitary controls, the irrigation of vegetables eaten raw with reclaimed water, no matter how well treated, remains prohibited in Jordan.

In Mexico, implementation of waste-water treatment (but not necessarily its use) has been mandated by federal environmental quality regulations. While waste-water use in agriculture is a common practice, particularly in Mexico's vast arid and semi-arid areas, it is mostly practised informally with the result that planned treatment for use in agriculture is not common. Instead, municipal water boards that bear the cost of treatment prefer to seek paying customers for treated waste-water, particularly golf courses, urban green spaces, etc.

Just how prevalent waste-water irrigation is today is a matter of conjecture; no sound, verifiable data exist. Earlier approximations by Scott, based on figures for sewage generated, treatment capacity installed, assumptions of the proportion of peri-urban areas without waste-water demand for agriculture (e.g. coastal cities, etc.), freshwater mixing ratio, and annual irrigation depths, placed the area at 20 million ha of irrigation using raw or partially diluted waste-water. Since the release of this first-cut estimate, the reactions have been multiple that:

1. The 20 million ha figure is an overestimation of 'raw sewage irrigation' given that it includes areas irrigated with partially diluted waste-water

2. Waste-water irrigation is not important enough a phenomenon to warrant resources for research and management
3. The magnitude of the problem is significantly greater than that implied by the 20 million ha estimate
4. Isolated case studies barely scratch the surface and indeed irrigation using waste-water or seriously polluted water is pervasive and represents a major concern.

Clearly there is a need to establish and apply a verifiable method for determining the prevalence of waste-water irrigation. As an important first step in this direction, van der Hoek presents a typology. Raschid-Sally et al. and Cornish and Kielen present assessments at the country level with estimates of 9,000 ha for Vietnam and 11,900 ha for Ghana. Ensink et al. estimate that 32,500 ha are irrigated with waste-water in Pakistan. These results are based on a typological definition of undiluted waste-water, i.e. 'end-of-pipe' sewage irrigation, which does not account for irrigation using water polluted with waste-water, that poses many of the same risks and management challenges.

Van der Hoek's typology includes marginal quality water, i.e. polluted surface water; however, country estimates have tended to focus on undiluted waste-water irrigation, suggesting that 20 million ha is an over-estimation of the global extent of the practice. It is important to recognise, however, that improved estimates of global waste-water irrigation would need to account for a number of countries with rapidly growing cities and large national irrigation sectors including particularly China, Egypt, India, Indonesia, Iran, Mexico, and Pakistan. This does not detract from the importance of waste-water irrigation or the difficulty of the management challenges in other countries or regions. Multiple complementary factors drive the increased use of waste-water in agriculture. Water scarcity, reliability of waste-water supply, lack of alternative water sources, livelihood and economic dependence, proximity to markets, and nutrient value all play an important role. Water scarcity and reliability of waste-water supply are crucial. That farmers have few alternative water sources may be true where waste-water is mixed with freshwater; however, in water-scarce regions, waste-water is invariably the only source.

Interestingly in some cases, as in Pakistan where canal irrigation water is available, although with reliability and supply constraints particularly in the tail-end reaches of the irrigation systems, many farmers convert to wastewater by choice. Livelihood dependence for poor farmers remains the single most important socioeconomic driver of the practice, yet it is misleading to assume that all waste-water farmers are poor. Indeed, larger, commercial-scale farmers have made inroads and may compete with small-scale farmers for waste-water as well as for markets.

Additionally, because of the market orientation of much waste-water agriculture in urban and peri-urban contexts, it absorbs significant labour, much of it female. Finally, while most farmers acknowledge the nutrient value of waste-water this appears to be a secondary driver, i.e. the scarcity or poor quality (usually salinity) of alternative sources is generally more important.

Waste-water irrigation will remain consigned to informal practice and as a result management approaches must start at the informal or semi-formal level. While the use and livelihood dependence on waste-water in African cities is not entirely dissimilar, it is hypothesised that social relations and land tenure issues related to state or communal ownership of land may not result in the same formalisation of waste-water irrigation in urban and peri-urban agriculture as seen in Asian cities.

Uni-dimensional management solutions for waste-water irrigation that employ exclusively technical (treatment) or regulatory (bans, crop restrictions, etc.) approaches have generally been inadequate. In isolation neither fully takes account of the multiple drivers of the process, nor the need for integrated management solutions. Realistic and effective management approaches rarely hold up technical or regulatory approaches as the complete solution, but instead seek to apply these in an integrated way. The more difficult question, particularly in the context of weak regulatory implementation, lies in the multiple - often competing - needs to secure livelihoods based on waste-water irrigation on the one hand, and public health and environmental protection imperatives on the other.

Health and Environmental Quality

The single most important rationale for more stringent control over waste-water use in agriculture is the risk posed to human health (of irrigators, consumers of produce, and the general public) and to the environment. Guidelines for waste-water use and standards for water quality matched to particular end uses have been developed and applied with varying degrees of success. Two sets of guidelines that aim to protect human health under conditions of planned reuse of treated waste-water - those set out by the World Health Organization and the United States Environmental Protection Agency (USEPA) - have raised considerable controversy in particular with respect to their feasibility and applicability in different developing country contexts.

Fattal et al. estimate that the cost of treating raw sewage used for direct irrigation to meet the current WHO microbial guideline of 103 faecal coliforms/100 ml is approximately US$125 per case of infection (of hepatitis, rotavirus, cholera, or typhoid) prevented. By comparison, the incremental cost of further treating waste-water from the WHO to the USEPA microbial guideline is estimated to be US$450,000 per case of infection prevented. Developing and applying pragmatic guidelines based on managed risk or acceptable risk instead of 'no risk' criteria must be the approach adopted.

There are two primary constraints to the adoption of any set of guidelines: firstly infrastructure, operation and maintenance, and the associated investment and recurring costs that are required to handle or treat waste-water to the quality levels stipulated in the guidelines, and secondly regulatory enforcement to ensure compliance with required practice on the part of water authorities, those discharging waste-water, and those handling and using waste-water.

Invariably the infrastructure issue is seen as the principal challenge, so that much of the debate is centred on waste-water treatment plants, their design, cost of operation, maintenance, etc. The assumption appears to be that with adequate technical control, the need to limit waste-water discharge and subsequent use is sufficiently minimised. This places ultimate responsibility for guidelines compliance on urban development authorities who control the finance of waste-water infrastructure and on waste-water treatment plant operators. Yet in the case of planned reuse there are larger institutional issues that permit (or impede) the implementation of waste-water use programmes, of which guidelines may be an important component.

In developing-countries, use of waste-water is an unplanned activity, and authorities tend to view the responsibility of regulating its use as a burden. In the absence of resources for treatment infrastructure and regulatory control, the guidelines proposed by the WHO, while relevant in a planned reuse context, are relegated to the status of targets (usually unachievable) instead of norms for practice.

REFERENCES

Asano T. and Tchobanoglous G. (1987) *Municipal wastewater treatment and effluent utilization for irrigation.* Paper prepared for the Land and Water Development Division, FAO, Rome.

Bartone C.R. (1986) *Introduction to wastewater irrigation: extent and approach.* Third Annual World Bank Irrigation and Drainage Seminar, Annapolis, Maryland. The World Bank, Washington DC.

Department of the Environment. (1989) C*ode of Practice for Agricultural Use of Sewage Sludge.* Her Majesty's Stationery Office, London.

FAO. (1985) *Water quality for agriculture.* R.S. Ayers and D.W. Westcot. Irrigation and Drainage Paper 29 Rev. 1. FAO, Rome. 174 p.

Mara D.D. (1976) *Sewage Treatment in Hot Climates. J*ohn Wiley, London.

Pescod M.B. (1987) *The quality of effluent for reuse in irrigation.* Paper prepared for the Land and Water Development Division, FAO, Rome.

Bibliography

Al-Labadi, M., "Water Harvesting in Jordan- Existing and Potential Systems", *In: FAO, Water Harvesting For Improved Agricultural Production*, Expert Consultation, Cairo, Egypt 21-25 Nov. 1993.

Allen, RG,L.S. Pereira, Dr. Raes, M. Smith, *Crop Evapotranspiration (Guidelines for Computing Crop Water Requirement).*, FAO Irrigation and Drainage Paper No.56. 1998.

Asano T. and Tchobanoglous G. (1987) *Municipal wastewater treatment and effluent utilization for irrigation.* Paper prepared for the Land and Water Development Division, FAO, Rome.

Bartone C.R. (1986) *Introduction to wastewater irrigation: extent and approach.* Third Annual World Bank Irrigation and Drainage Seminar, Annapolis, Maryland. The World Bank, Washington DC.

Boers T.M. and Ben-Asher J. *Harvesting water in the desert.* In: Annual Report 1979, International Institute for Land Reclamation and Improvement (ILRI) Wageningen, Netherlands. 1979

Bonsu, M., "Organic residues for less erosion and more grain in Ghana," In: Soil Erosion and Conservation, El-Swaify, Moldenhauer and Lo (eds)., *Soil Cons. Soc. Amer.*, Ankeny, Iowa. 1985.

Bruins, H. J., Evenari, M. and Nessler, U., *Rainwater harvesting agriculture for food production in arid zones: The challenge of the African famine*, Appl. Geogr. 6:13-33, 1986.

Cannell G.H. (ed). *Proceedings of an International Symposium on Rainfed Agriculture in Semi-Arid Regions,* Riverside, California, April 1977. 1977

Chartres, C. and Varma, S. *Out of water. From Abundance to Scarcity and How to Solve the World's Water Problems* FT Press (USA), 2010.

Cullis Adrian, Pacey Arnold, *Rain Water Harvesting,* London, Intermediate Technology Publisher, 1991.

Department of the Environment. (1989) *Code of Practice for Agricultural Use of Sewage Sludge.* Her Majesty's Stationery Office, London.

EPA. "How to Conserve Water and Use It Effectively". Washington, DC. Retrieved 2010-02-03.

FAO. (1985) *Water quality for agriculture.* R.S. Ayers and D.W. Westcot. Irrigation and Drainage Paper 29 Rev. 1. FAO, Rome. 174 p.

Geerts, S.; Raes, D. "Deficit irrigation as an on-farm strategy to maximize crop water productivity in dry areas". *Agric. Water Manage* 96 (9): 1275–1284. 2009.

Ladda, Brijgopal, *Planning and Financing Urban Water Supply: A Case Study of Ahmedabad,* Ahmedabad, CEPT, 1995.

Mara D.D. (1976) *Sewage Treatment in Hot Climates.* John Wiley, London.

Matlock W.G. and Dutt G.R. *A Primer on Water Harvesting and Run-off Farming.* 2nd Ed. Agric. Eng. Dept., College of Agriculture, University of Arizona. 1986

Melden , D., *Water for food, water for life; a Comprehensive Assessment of Water Management in Agriculture;* International Water Management Institute (IWMI). 2007.

Molden, D. (Ed). *Water for food, Water for life: A Comprehensive Assessment of Water Management in Agriculture.* Earthscan/IWMI, 2007.

Nairizi, I,R. Redzeweski, *Effects of Dated Soil Moisture Stress on Crop Yields;* Irrigation Engineering Research Group, Department of Civil Engineering University of Southampton, 1976.

National Academy of Sciences. *More Water for Arid Lands. Board of Science and Technology for International Development,* Commission on International Relations. NAS, Washington DC. 1974

Pacey A, with Cullis A. *Rainwater Harvesting - the Collection of Rainfall and Run-off in Rural Areas.* Intermediate Technology Publications, London. 1986

Pearce, Fred *When the Rivers Run Dry: Water—The Defining Crisis of the Twenty-First Century* Beacon Press, 2006.

Pescod M.B. (1987) *The quality of effluent for reuse in irrigation.* Paper prepared for the Land and Water Development Division, FAO, Rome.

Pimentel, Berger, *et al.* "Water resources: agricultural and environmental issues". *BioScience* 54 (10): 909. 2004.

Redzeweski, J.R, S. Nairizi, I*rrigation Planning Based on Water Deficits*; Water Resources Bulletin American Water Resources association, 1979.

Siegert, K., "Introduction to Water Harvesting. Some Basic Principles for Planning", *Design and Monitoring. In: FAO, Water Harvesting For Improved Agricultural Production*, Expert Consultation, Cairo, Egypt 21-25 Nov. 1993.

Stewart B.A., Unger P.W. and Jones O.R., "Soil and water conservation in semi-arid regions," In: Soil Erosion and Conservation, Ei-Swaify S.A., Moldenhauer W.C. and Lo A. (eds), *Soil Cons. Soc. Amer.*, Ankeny, Iowa, 1985.

The World Bank. *Sustaining Water for All in a Changing Climate.* The International Bank for Reconstruction and Development. 2010.

U.S. Environmental Protection Agency (EPA) Cases in Water Conservation (Report). Retrieved 2010-02-02.

UNEP. *Rain and Stormwater Harvesting in Rural Areas*. Ed. UNEP, Tycooly International, Dublin. 1983.

UNESCO; *Water in a Changing World,* the United Nation World Water Development Report. 2009.

Vickers, Amy (2002). *Water Use and Conservation.* Amherst, MA: water plow Press. 2002.

Walker P.J., "Cropping and soil conservation in semi-arid western New South Wales (Australia)," *J. Soil Conservation Service of New South Wales* 38(2): 49-56, July 1982.

Willcocks T.J., "Tillage requirements in relation to soil type in semi-arid rainfed agriculture," *J. Agric. Eng. Res.*, 30: 327-336, 1984.

Index